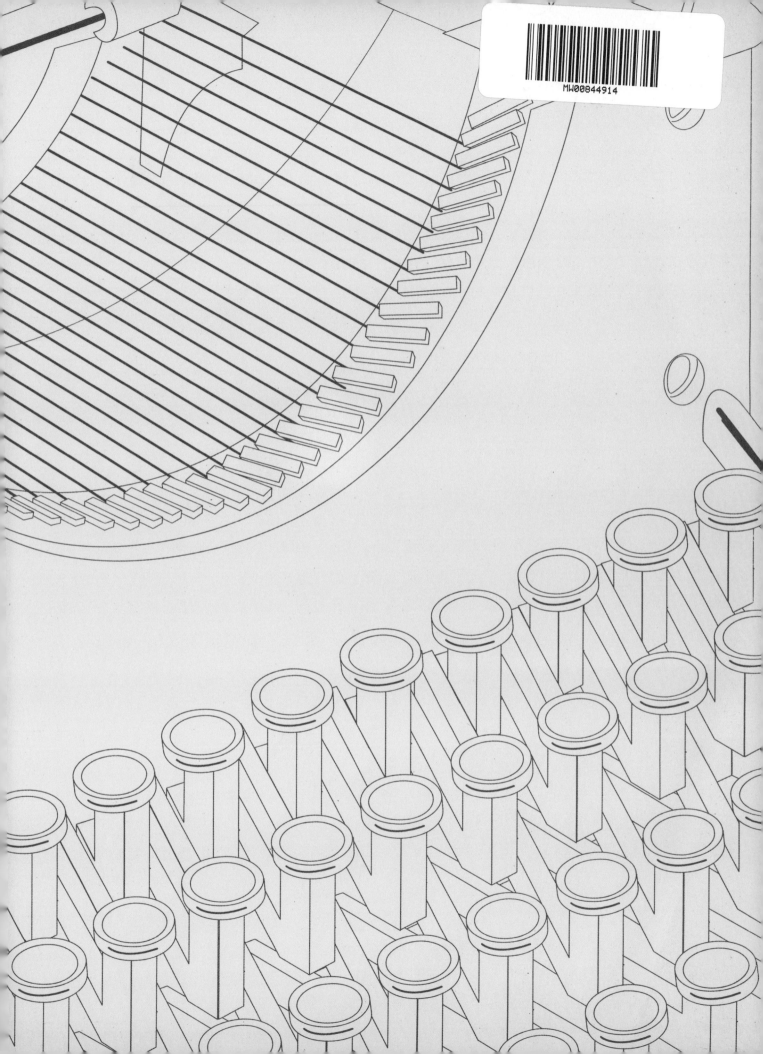

ANTIQUE TYPEWRITERS
From Creed *to* QWERTY

'*A machine to die for*': Malling Hansen's Skrivekugle (Courtesy of Sotheby's, London)

ANTIQUE TYPEWRITERS

From Creed *to* QWERTY

Michael Adler

77 Lower Valley Road, Atglen, PA 19310

to my littlie GERRIE

Printed in China

ISBN: 0-7643-0132-2

Book Design by Audrey L. Whiteside

Library of Congress Cataloging-in-Publication Data

Adler, Michael H.
   Antique typewriters, from Creed to QWERTY/Michael Adler.
      p.   cm.
   Includes bibliographical references and index.
   ISBN 0-7643-0132-2
   1. Typewriters--History. 2. Typewriters--Collectors and collecting. I. Title.
Z49.A1A32   1997
681'.61'09--dc20   96-33429
                                      CIP

Published by Schiffer Publishing, Ltd.
77 Lower Valley Road
Atglen, PA 19310
Phone: (610) 593-1777
Fax: (610) 593-2002
Please write for a free catalog.
This book may be purchased from the publisher.
Please include $2.95 for shipping.
Try your bookstore first.

We are interested in hearing from authors
with book ideas on related subjects.

Contents

Acknowledgments

No book is ever written without a great deal of help from family, friends and associates—I am genuinely grateful to them all, collectively and individually, for their assistance over the years.

I am particularly indebted, yet again, to that tireless and uncomplaining researcher Barbara Spadaccini-Day for her many contributions, as well as to Richard Brown and to the late Maureen Dowd for their help with the irksome task of reading the manuscript.

My thanks also go out literally to the four corners of the earth, to the indefatigable GiGi Konwin of This Olde Office, Cathedral City, California, for her personal and professional altruism and enthusiasm; to Tom Fitzgerald in Philadelphia; to John Foster in New Zealand and to Professor Yamada Hisao in Japan for their individual contributions; to Bernard Williams in England; to the late Paul Lippman in New York and the late Dan Post in California; to Sandy Sellers in Ontario, Canada; and to John Lewis Sr. in Albuquerque, New Mexico; to Catherine Dessau in Rio de Janeiro, Brazil; as well as to Mara Miniati and Franca Principe of the Istituto e Museo di Storia della Scienza in Florence, Italy; to André Desvallées of the Conservatoire National des Arts et Metiers in Paris, France; to Margaret Hobbie of the DeWitt Historical Society in Ithaca, New York; and to Sally Grover of The Royal Society, London. And of course, last but not least, to Christopher Proudfoot of Christie's South Kensington, London; to Jon Baddeley of Sotheby's, London; to Diana Kay of Phillips, London; and to Alex Crum Ewing of Bonhams, London. My thanks again to you all.

1. Sholes & Glidden Type Writer (Courtesy of Phillips, London)

Introduction

It was not so long ago that only the most intrepid of men would admit publicly to collecting old typewriters. One would have to steel oneself for the inevitable hilarity which would greet such a confession. Benign incredulity was the best one could hope for from one's bemused audience. Downright derision was what one usually had to face.

Not any more. A collecting field tends to acquire respectability and public acceptance only after an authoritative book on the subject has legitimised what was previously deemed to be nothing more than individual folly and eccentricity. My first book, *The Writing Machine*, appears to have gone some considerable way towards performing that function. [1]

Certainly, when I look back over considerably more than a quarter of a century of involvement in collecting and studying old typewriters, I cannot help but be dumbstruck by the changes which have occurred. In the 'olden days' there were countless machines but virtually no collectors. There was a chap in France and one in Italy, one in Switzerland, a couple in England and a few more in the States. That was all. And there were machines everywhere, absolutely everywhere. Nobody wanted them or knew anything about them. These days, if you walk into a shop inquiring innocently after old typewriters, you are just as likely to be told, 'No, we have no Sholes and Gliddens today. Sorry, and we sold our last Typograph only yesterday' **(1, 2)**. Everyone, or so it seems, has

2. Hughes' Typograph (Courtesy of J. G. Foster Collection)

suddenly become an expert and collecting typewriters is at last a respectable eccentricity.

So what is it, I have often been asked, that turns an otherwise normal person into a typewriter enthusiast? There is no ready answer. It helps no doubt if you are vigorous and healthy. You also need a little spare cash, and a great deal of spare time. But above all you need passion. Hot blood must course in your veins and logic and reason must only rarely be permitted to intrude. If you stop to worry about straining your back as you run after a bus with a Kanzler **(3)** under your arm, then you ought to be collecting stamps or cigarette cards. If you can hear the word *Skrivekugle* **(frontispiece, 4 et al.)** without instant adrenal exhaustion, then typewriter collecting is not for you.

Other than that there are no pre-requisites. Some collectors begin in their early teens, while others start only from the proceeds of their pensions. Some mix machines with children, others have machines instead of children. Some marry—many times—while others remain single. Some go to their neighbours' to bathe because their own tubs are full of soaking typewriter parts. Others simply never bother to bathe at all.

There are old timers who are true to their hobby for the better part of their adult lives. Others tire of it or become disillusioned and sell their collections to a new generation of enthusiasts. A few, sadly, have died and their collections are now dispersed. Some never manage to assemble more than a handful of machines, while others have private museums, which, for size and quality, are nothing short of spectacular. There are collectors rich enough to pay any price for a machine they want. Others are so poor that they can only buy a typewriter if they stop eating for as long as it takes to pay it off. A few finance their modest collections by the heart-breaking expedient of selling

3. Kanzler (Courtesy of Bernard Williams)

anything good that comes their way and buying cheaper, but naturally less desirable, items with the proceeds. Some boast about the world record prices they pay, while others boast that everything they buy is dirt cheap.

There are those who complain that they pay too much, and those who proclaim that they never pay anything at all. And sometimes, amazingly enough, that it is not all too far from the truth, either. An East Coast collector once offered to exchange one of his spare machines for one of mine. I had it professionally crated and shipped over to him from England, and in due course I received an enthusiastic letter from him acknowledging receipt and promising to send his machine by return. That was the last I heard from him and that was close to twenty years ago. I myself tried to contact him on various Transatlantic trips, and friends have also tried to intercede, but to no avail. He shows off my machine with great pride, to this day, but has never concluded the reciprocal part of the deal.

4. Malling Hansen's Skrivekugle (Courtesy of Christie's South Kensington, London)

Such an experience is a salutary lesson, although how one protects oneself against rogues like that is hard to imagine. Thankfully there are few of them around and it is gratifying to know that, by and large, despite the inevitable chatter and patter, typewriter collectors are a remarkably honourable bunch. They may bicker and squabble every so often but they also know how to kiss and make up. However, some hilarious anecdotes can result from the inter-relationships between collectors. A friend once exchanged one of his machines for a Hall and when the overseas parcel eventually arrived, it struck him as being a little small. On opening it, he discovered that it contained a bunch of nuts, bolts, screws and assorted bits of metal of varying shapes and sizes, all of them apparently Hall-esque in appearance…but worse even than that: there was no case at all, which (as *cognoscenti* well know) is essential in a Hall, because the machine does not just pop in and out of its case like most others do but comes screwed into it **(5)**. So where's the case, cries the friend in amazement over the phone. Oh, replies the other party nonchalantly, I never told you there was a *case*! (He did eventually agree to take it back, though, so all was well in the end).

The one trait that all typewriter collectors share is that they are all, to a greater or lesser degree, obsessive. Is it a fault? It probably is, although I am in no position to criticize because I was worse than obsessive myself, in my time—I was positively possessed! The passion was demonic, and what made it worse was that the objects provoking the passion were everywhere to be found in those early days—all you had to do was tolerate the occasional ridicule. I recall the time I stopped off at a likely-looking junk yard in provincial France, where perhaps the most desirable object was half a rusty 1930s ice box. When I asked the gorilla in charge whether he had any old typewriters, he quite literally fell apart, bellowing with laughter and slapping his substantial thighs in uncontrolled derision. 'Now it's *des machines à écrire* they're asking for! Whatever will they want next?' he gasped, rhetorically.

I was not sure what they would want next, quite frankly, unless it was their edible entrails of oxen, perhaps, or their gourmet gastropods, so I simply drove off duly chastised, and in something of a hurry I might add, since I allowed myself the luxury of revealing to the oaf a few salacious facts he never suspected about his mother.

But that sort of incident was rare. More often than not, one was welcomed with open arms by men desperate to liberate valuable space long wasted on storing worthless old typewriters.

When I was in my early twenties I hitchhiked around South America from Brazil through Uruguay, Argentina, and Chile to Peru where I bartered a pair of boots for a dug-out canoe from a local Indian and paddled it down the Ucayali and Amazon Rivers and up the Rio Negro to Manaus. It seemed a good idea at the time and I survived those thousands of miles of uninterrupted nightmare only because God has a unique way of protecting idiots even from themselves. I thought then that nothing I would ever have to face would match that experience for sheer effort and agony.

5.   Hall Type Writer (Courtesy of J. G. Foster Collection)

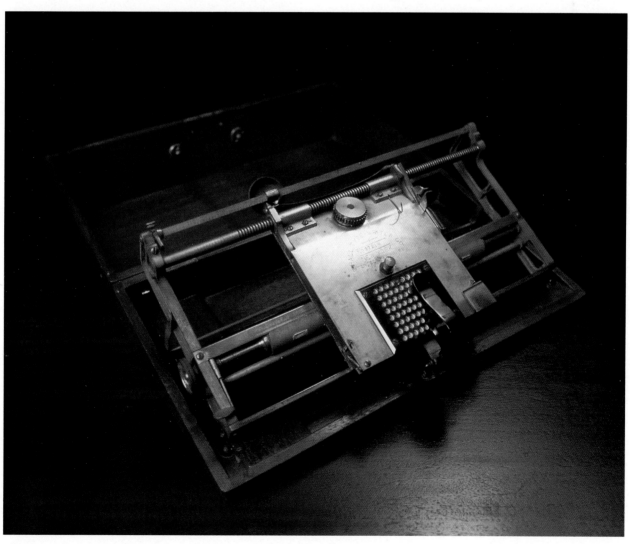

Yet it paled into insignificance when compared to the things I was later to do in my search for typewriters.

I was living in Rome by then, in the heart of the old city, and I was as besotted by Rome in the 60s as was Henry Miller by Paris in the 30s. The Sunday market at Porta Portese was just across the Tiber. I came back from my first visit clutching a little Frolio, which I thought was so wonderful I started using it for typing my correspondence (6 [r]). Probably every typewriter collector has a similar story to tell. Not that I had much time for typing, however, because I went mad looking for more machines.

tween, we visited minor towns and cities, museums, libraries, and patent offices. And we also wrote letters…tens of thousands of letters.

Interesting patterns quickly began to emerge. All the world's recorded Suns (8), at the time, came from Madrid and Barcelona. To get at them, you were invariably taken to basements or barns where you had to excavate through successive strata of Continentals and Mercedes (typewriters, that is), Monarchs (193) and Underwoods (75). All the Columbia Type Writers (9, 73, 137) came from London—I once bought two at Portobello on the same

6. Junior (l), Frolio (r) (Courtesy of Bernard Williams)

In retrospect, I feel privileged to have been able to enjoy all those fertile and productive years of my life while old typewriters were forgotten and neglected artifacts. It has been said that by and large a man achieves everything he is going to achieve between the ages of 30 and 40—what he does before that decade is essentially preparation, and what he does afterwards is nothing but more of the same. If this maxim is true, then indeed I must have been blessed by fate! My wife and I would often drive from Rome to London to be at Bermondsey Market for 5am on Friday and immediately back via the Marché aux Puces in Paris, returning just in time to Porta Portese in Rome at sunrise on Sunday. Sometimes we would substitute London's Portobello on the Saturday. Once a month we went to the market at Arezzo—and every little town along the way, too, because every one had an antique shop or two with old typewriters stacked away somewhere in the back. At times we would sacrifice some of these regular visits to go by back roads to Barcelona, Madrid, Vienna, Amsterdam, Zurich or even Prague (which yielded a Helios Klimax [7]), or Budapest (which yielded nothing but goulash). Several times a year we took trips further afield to the New World, from New York to Buenos Aires. In be-

morning. Foxes (10) came from England too and Rofas (11) from Holland. Sholes and Gliddens, such as there were, came from the States, as one would expect (rear cover, 12 et al.). In those days you could count on the fingers of one hand the total known number of S&G machines surviving outside museums. Now there are upwards of fifty recorded, although it has taken some thirty years for them to emerge from the woodwork. The first S&G I ever bought was in New York, and I cradled the precious bundle on my lap all the way back across the Atlantic. I had to pay a cool thousand dollars for it too, which was the price you paid for a new car

7. Helios Klimax (Courtesy of Bernard Williams)

8. Sun No. 2 (Courtesy of This Olde Office, Cathedral City, California)

9. Columbia Type Writer No. 2

10. Fox (**l**), Blick 90 (**r**) (Courtesy of Bernard Williams)

12. Sholes & Glidden Type Writer
(Courtesy of Smithsonian
Institution)

11. Rofa (Courtesy of Bernard Williams)

in those days. I had no choice—it was the only available one I had ever come across and might possibly have been the only one left in the world, for all I knew. I justified it by writing it off against all the machines I had managed to pick up for next to nothing. You had to do that sometimes, if for no other reason than for the sake of domestic peace and harmony.

Of course while all this was going on, I was beginning to accumulate quite an extensive library and a considerable store of knowledge and experience to go with it. Not a lot of accurate research had been done on the subject up to that time and so much contentious and conflicting nonsense had been written that it eventually seemed like a good idea to set the record straight once and for all.

My first book, *The Writing Machine*, was the result. [1] In it I made a point of cross-referencing all this conflicting information to its sources and correcting it, and having done so once, there would seem to be little point in doing it again, so I have not repeated the exercise in the following pages and serious students will have to refer to that book for the historical record of previous background mythology which that book rendered obsolete. Nor have I thought it necessary once again to re-iterate the rather detailed explanations and diagrams of the classification of early typewriters according to their typing actions which I proposed in that book, because these have meanwhile become standard terminology, used and understood by all both within and without the typewriter collecting fraternity.

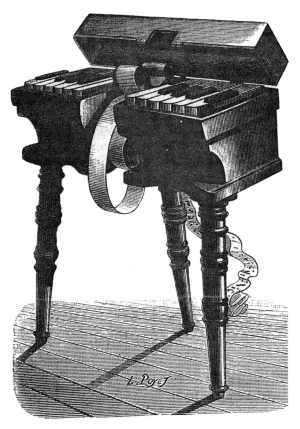

13. Michela[57]

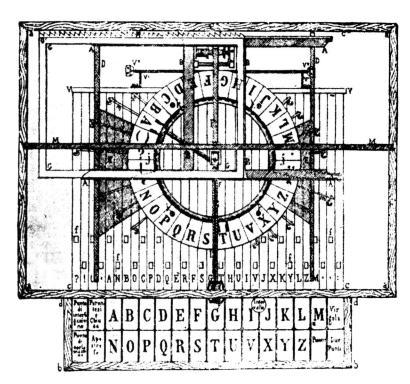

15. Ravizza's Cembalo Scrivano (Courtesy of Italian Patent Office)

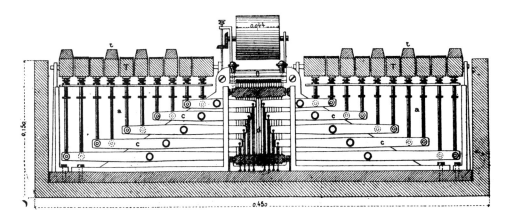

14. Michela[57]

17. Edison's Mimeograph (Courtesy of Bernard Williams)

16. Foucauld's Raphigraphe[27]

18. Hammonia (Courtesy of Bernard Williams)

complete and often inaccurate. [62] Among the newer books, Beeching offers some interesting information on more recent history and development,[7] while Tilghman Richards offers little more than a minimal outline. [88] The Milwaukee Museum booklet by Herrl on the other hand is an excellent reference although it too needs some correcting. [42] Lippman's recent book on American machines is a useful reference, too; other older American sources can be so outrageously partial, however, that they ought to carry a health warning. [60] Oden, for example, states that 'all the best typewriters that have been produced have been built in America,' an assertion which Europeans might well be tempted to dispute. [72] As for German-language sources apart from Martin,[63] Granichstaedten-Czerva's and Krcal's accounts of Mitterhofer are interesting but they too are flawed by bias,[36, 54] and the Baggenstos booklet is all but worthless. [6] The same may also be said of all the early Italian literature which is devoted almost exclusively to the Conti and Ravizza sagas, while French sources are notoriously inaccurate. [26, 27, 28, 80, 81, 84] Poor Turri was almost completely ignored by them all, even by the Italians, until I revealed his achievements in my first book—I wonder what his countrymen would have made of him, had they but realised.

Regrettably for the Italian nation, however, I have since succeeded in excavating ever deeper into typewriter pre-history in the course of the past twenty years and have been fortunate enough to be rewarded by unearthing the previously unknown information about Reverend Creed and his musical writing machine, and this discovery takes our history back from the early 19th to the mid-18th century.

As for the future, who knows what great and exciting finds lie in store for us. Fish from the bowels of the Indian Ocean are not the only species presumed extinct which could burst out upon a startled world. Somewhere out there in an attic or cellar is a Michela Stenograph (13, 14) or a Cembalo Scrivano (15) or a Foucauld Raphigraphe (16), to name but a few, and the enormous worldwide interest in these old machines will not only help to find them out but will also ensure that once found, they stay found forever more, adequately restored and preserved for all time. Money will not necessarily determine who finds them either; more than anything else it will probably be luck. One lady I know began her collection with an Edison Mimeograph (17), another with a Hammonia (18). After lifetimes without finding either, many seasoned collectors out there are only too aware how a perverse fate can so easily make grown men weep. Enthusiasts simply have to learn to cope with such injustice!

Nevertheless there are still early books which are well worth tracking down, since no one source can ever aspire to the inclusion of every known fact. Notable among these is the book by Ernst Martin which, despite its many mistakes and dreadful bias, remains one of the standard texts on the subject. [63] But *bias*? I love the sour grapes of the man's quite unsubstantiated claim that 'wahrscheinlich auch bei uns in der damaligen zeit, vielleicht schon früher…'—his thinly disguised distaste at having to admit that the first typewriter patent (Henry Mill) was English and not German was obviously a pill too bitter for him to swallow.

Bias is always a big problem and every nationality is guilty of gross distortion at some stage, in typewriter history as in every other field. This needs to be borne in mind when dealing with the available literature. The *Greater Soviet Encyclopaedia* for example—which at one time was the source of all knowledge to quite a significant portion of the world's population—still lauds Comrade Alissoff as the inventor of the typewriter. Fortunately, not all sources are as biased as this. Martin, as I have said, is a little too partial for my liking and in such details as relegating Turri to nothing more than a few lines on page 349, is by now more than a little dated. [63] Mares is patchy but worth reading, even though it is in-

But you can be lucky with lesser objects than a Hammonia, as I only recently discovered. The earliest writing machine which I can remember using is the penny-in-the-slot circular index device which often graced railway stations and arcades when I was a child. Any time I could get hold of a penny, which was not all that frequently in those days, I would treat myself to the thrill of embossing a usually incomplete message on the thin aluminum strip. The trick was to compose a text with precisely the right number of letters and spaces before the penny's worth ran out. No point wasting money on a message that was too short! It must have been either before the War or shortly after it, because during the War one would not have been permitted to 'waste' aluminum like that. I had presumed these machines to be extinct, like the aforementioned coelacanth, until I happened to stumble upon a badly neglected one not so long ago; just enough paint remained on protected areas to indicate its original color, and the restored example (only just recognizable in the poor accompanying illustration [19]) is once again in daily use in a museum collection. Maybe all the others were recycled during the War—into armaments, more likely than not.

I had never really considered these machines as typewriters when I was a child, but they certainly were. They were of circular index design and they embossed rather than printed with carbon paper or ribbons. However, plenty of historical typewriters had precisely those specifications and of course machines for the blind will perpetuate the embossing principle forever, as will those modern office gadgets which emboss thin strips of plastic. The old aluminum penny-in-the-slot embosser was therefore a kind of primeval dinosaurian Dymo! This is the beauty of old typewriter designs—unlike their creators or chroniclers, none of them is ever completely dead and buried.

Meanwhile the body of knowledge is constantly growing thanks largely to the enthusiasm of collectors in comparing specifications of their own machines. Sometimes the discussions and arguments must seem trivial to outsiders, but to the aficionado the smallest details can be of vital importance.

Take the Williams (**20 et al.**) for instance. Long a favourite with collectors because of its characteristic grasshopper typing action, the Williams went through some minor modifications which the manufacturer had considered too unimportant to specify. By the time he had introduced a machine which he labelled Model Two he had already marketed two major versions, each with significant differences, of his first model—although neither of these was ever labelled Model One. Co-operation between collectors around the world comparing specifications of the machines in their possession set the record straight.

20. Williams 2 (Courtesy of Christie's South Kensington, London)

19. The primeval 1930s penny-in-the-slot ancestor of the Dymo.

In fact, very few manufacturers actually marketed a machine which they labelled Model One from the start. Some, like Remington, did a little of this retrospectively; most others did not bother at all. The upper-case-only Columbia type wheel was the first model of that make, but was never named Model One. The single-wheel upper and lower case version was indeed called Model Two, but briefly between these two, a double-wheel variant also appeared **(21)**. Or take the Salter **(22)**. At the time I wrote *The Writing Machine*, as far as anyone knew the first model for some reason or other was called Improved Model 5. There was no existing record of an ordinary Model 5, nor indeed of versions 1 to 4. It was only over the years that we were able to piece together the probable explanation, which is that the numbers related to the various modified inking systems the manufacturer tried before he finally got it right. Co-operation between collectors can often pinpoint these changes fairly closely. When you find a Salter 6 serial number 6205 and compare it with the Salter Improved 6 number 7803, you have established that the model change occurred somewhere between those numbers. If another collector then produces a model 6 number 7698, you have immediately reduced from 1600 to just over 100 units the point at which that minor model change was introduced.

Does all this really matter? Certainly it matters. It matters a great deal. No book ever written represents the last word on a particular subject. Each one is but a stepping stone to the next and readers should pencil in their own additions and corrections wherever appropriate. Far from defacing a book, I believe that this actually enhances rather than diminishes its value. I have in my own library definitive horological tomes from past generations in which illustrious former owners have written their comments. I cherish these volumes above all others. They are dynamic and live on after their owners are long dead. Had the information been stored in a file or card index, it would have been lost. Filing cabinets and their contents together with all the other office paraphernalia rarely outlive their original owners, but books are different. Books are valuable and are passed down or sold off after their owners' deaths and are seldom, if ever, willfully destroyed. And they have that remarkable quality of finishing up in the hands of the very people who are most interested in their contents. Hopefully our books on typewriters will be no exception and will some day pass on our collecting passion to future generations.

—Michael Adler, Sussex.

21.   The 'double wheel' Columbia Type Writer (Courtesy of Bernard Williams)

22. Salter Improved No. 5 (Courtesy of Christie's South Kensington, London)

Chapter One—
First Generation:
The Age of Invention

Much Brazilian ink was spilt some decades ago promoting the candidacy of a priest called Azevedo as the father of the typewriter. This might come as something of a surprise to those who believe that Brazil is arguably a most unlikely venue for such a momentous historic event. Of course Azevedo in 1861 was not the inventor of *the* typewriter but was simply the inventor of *a* typewriter **(23)**, like many others before him and, had he been of US or British origin, he would hardly have warranted more than the most perfunctory mention. As it is, he was the only typewriter inventor the Brazilians had, so he became their national hero whose invention was 'stolen' and spirited overseas by a foreigner who promised to set up a manufacturing partnership, only to disappear without trace.

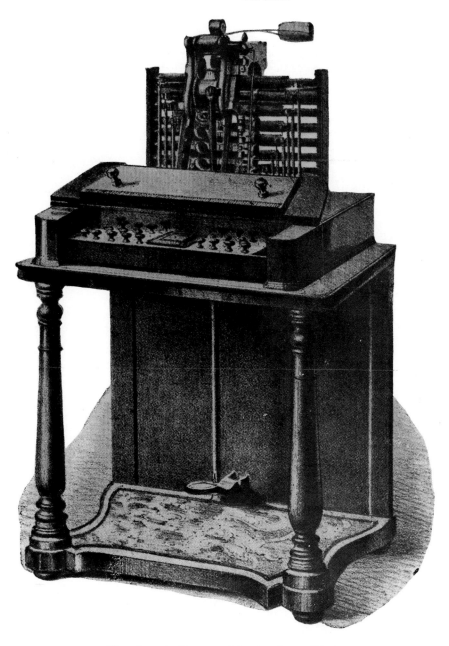

23. Azevedo (Courtesy of Brazilian Patent Office)

Ho hum! This is a familiar story in the annals of typewriter history. The Italians said it of Ravizza's inventions **(15)**, just as the Austrians said it of Mitterhofer's **(24)**. Alas, the facts do not bear out any of these allegations and whether the inventions were copied overseas is largely academic: none of these gentlemen qualifies as *the* first inventor, for the history of the typewriter began many decades earlier. It began not in the 19th but in the 18th century, during that wonderful age of enlightenment and discovery which provided the base from which the technological leap of the 19th century was launched. Science and technology, invention and discovery were partners with developing industry and commerce in this exciting era, all working together to create the new philosophical and psychological climate in which the great Industrial Revolution was to germinate.

The first recorded efforts at developing an instrument which was to become the typewriter are to be found here. The 18th century was a mere fourteen years old when an Englishman by the name of Henry Mill was granted a patent for a device capable of printing letters 'singly or progressively one after another as in writing.' Now this was undoubtedly a typewriter of some kind, and if Fate had been kinder we might have been able to point to an instrument, religiously tended in a shrine in some great museum and say, 'This is where it all began.' But alas, in Mill's day it was still possible to patent an idea or a project without having to provide any other details and we may therefore never know whether Mill actually made or even designed the instrument with which he has so tantalised future generations.

This, however, is the first recorded sighting of the idea of the typewriter and others were soon to follow. Hints of earlier efforts are indeed to be found, but they were all second or third hand reports and no confirmation has ever been located.[1] Until such evidence is offered, these earlier accounts must be dismissed as nothing more than hearsay.

But even if there is no extant record that Mill actually made a machine or left a drawing of his design we do not have long to wait because within a few decades another man appeared on the scene who not only had the bright idea of a writing machine but who left details and drawings to substantiate it.

A great deal of confusion has surrounded this matter, and conflicting claims have only succeeded in adding to the muddle. At the time I researched and wrote *The Writing Machine* I was still not able to establish the precise identity of a nebulous reference to a man allegedly called 'Cred' who was supposed to have invented a machine to be attached to an instrument like a harpsichord for recording a musical score on paper as it was being played.[63] Subsequent research has since enabled me to establish the identity of this man as one Reverend Creed who died some time in or before 1747. I say 'in or before' because a letter from a John Freke written in that year and reported in the Philosophical Transactions of the Royal Society, London, refers to him as the 'late' Reverend Creed.[82]

24. Mitterhofer (Courtesy of Dresden Technical University, Dresden, Germany)

I think the inclosed Paper is the Effect of great Ingenuity and much Thought,' wrote Mr Freke, 'and as the Subject-Matter of it may tend to give great Improvement and Pleasure to many, not only in our own Country, but every-where, I hope my presenting it may not be thought improper that it may thereby be printed and published to the world.

'It was invented and written by Mr Creed, a Clergyman, who was esteemed, by those who knew him, to be a Man well acquainted with all kinds of mathematical Knowledge. It was sent me by a Gentleman of very distinguished Merit and Worth; if therefore from hence this Paper shall be thought proper to be published in the Philosophical Transactions. (sic) It will prevent its being lost to Mankind. [82]

By the time Freke wrote that report, the Reverend Creed was long dead and buried, so that the mention of it was already once removed. Later generations shrouded this momentous historical event in even more confusion. Martin, as has been mentioned, calls him 'Cred' and incorrectly labels his invention a Melograph,[63] while Chapuis gets the spelling right but calls the invention a Megalophone.[16] Neither of these names is mentioned in original sources and their authenticity is questionable.

No such confusion, however, surrounds the machine itself **(25)** nor the manner of its use. The purpose was

...the Possibility of making a Machine that shall write Extempore Voluntaries, or other pieces of Music, as fast as any Master shall be able to play them upon an Organ, Harpsichord etc. and that in a Character more natural and intelligible, and more expressive of all the Varieties those Instruments are capable of exhibiting, than the Character now in use. [82]

This invention is described in greater detail in Chapter Five but briefly, it consisted of a cylinder turned at a precise speed (one inch per second is mentioned) 'by the Application of a circulating, not a vibrating, Pendulum' (that is, a balance wheel or a fly, perhaps) fitted beneath the keyboard of the musical instrument. Nail points or 'pencils' fitted beneath the keys came into contact with paper from an endless roll, producing longer or shorter strokes depending upon the length of time the individual keys were depressed.

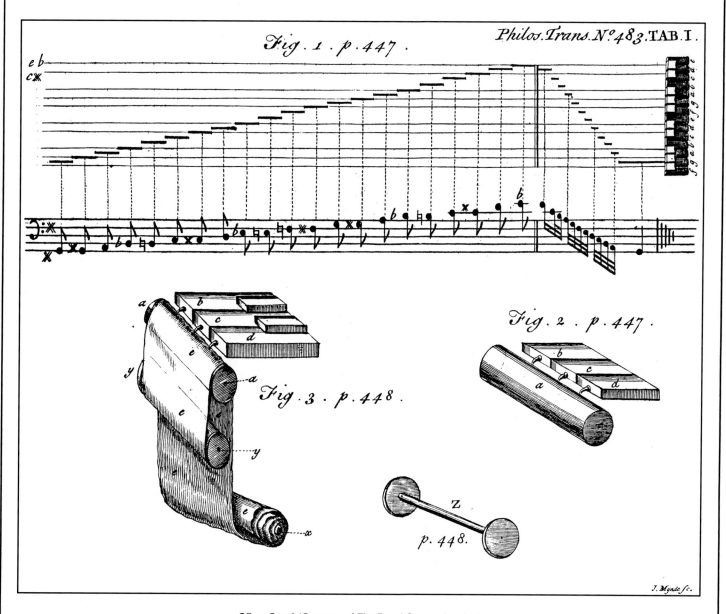

25. Creed (Courtesy of The Royal Society, London)

This would appear to be conclusive and ought to be sufficient to establish Creed beyond dispute as the inventor of the writing machine…except that a German by the name of Johann Friedrich Unger is also recorded as the inventor of an instrument for writing music in a similar manner, but using ink for the purpose instead. We are in fact dealing with the same machine and my research into these two gentlemen leads me to believe, beyond reasonable doubt, that the design was originally Creed's and that Unger later claimed it. The sequence of events would appear to be that Creed invented his machine in or before 1747 when it was first made public. Unger, using information from the Philosophical Transactions or possibly working completely independently on an identical invention (and this sort of coincidence has happened more than once in the course of history), first revealed his invention two years later in 1749 and then again in 1752, claiming, however, that he had completed it back in 1745 but without providing any corroborating evidence in support of this contention.

The invention itself is not in dispute and as we shall see later the same idea was proposed many times over in subsequent generations. However, it would be foolish indeed to accept an inventor's unsubstantiated claim when it comes to the dating of his invention. The fact is that the first public record of Unger's machine was contained in a letter he wrote to the Berlin Academy in 1749, by which time Creed was dead, and hardly in a position to dispute priority.

So Creed's would appear to have been the first writing machine in history, then—Henry Mill, remember, had left no indication of what he had in mind when he applied for his patent some thirty years previously.

Creed's idea made quite an impact, and not only on Unger. It surfaced time and again in different parts of the world in the century or so after his death, and it is easy to see why. One has only to observe the dexterity with which a pianist manipulates the keys of a piano to realise how well such an arrangement might lend itself to typing, be it of musical notes or of the letters of the alphabet. Creed and Unger were later joined by another German called Hohlfeld[63] or Holfeld[16] in 1771 or so **(26)**. Some refinements are evident in this later invention, for the roll of paper no longer needed to be the same width as the keyboard, which represented quite an improvement (logistical if not necessarily inspirational), but the idea was the same as Creed's, merely resurrected a quarter of a century later.

Another interesting development which dates from the same period, appearing time and again over the years almost to the present day, is represented by a device for making several originals of a handwritten message simultaneously. These pantographs are not really machines in the strictest sense and fall outside the scope of this book, but no collector I have ever met would pass one by if it came his way, so they are worthy of at least a mention. They are really nothing more than a frame holding two or more quills or pencils in such a manner that by writing with one of them, the others simultaneously record the identical motions on separate sheets of paper…in theory, if not always in practise.

Of all such devices, the most desirable is surely an example beautifully made by J. H. Farthing of Cornhill in the late 18th or early 19th century for the inventor Marc Isambard Brunel. He managed to patent the design on 17th January 1799, despite the fact that it was already well known and had been documented for more than thirty years.

Count Leopold Josef von Neipperg appears to have been the first with his Copiste Secret of 1762 and later examples offered no significant improvements or advantages. Why 'secret,' one might ask? Presumably because it permitted the writer to make copies of his document without the intervention of a scribe or secretary. On the other hand, just how practical such a device might have been is

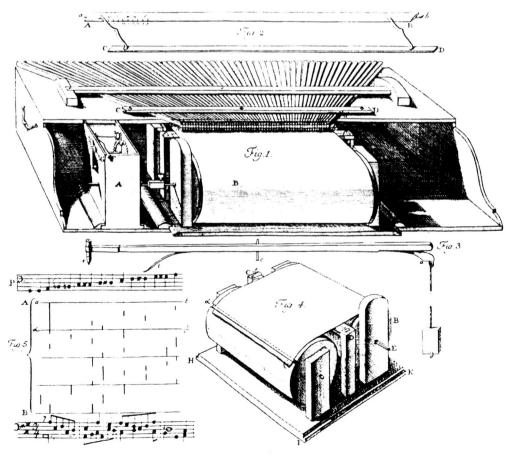

26. Hohlfeld[63]

highly questionable but is easily put to the test: merely fit two or more pencils or pens (not to mention quills!) onto a common handle in the manner of tines on the end of a garden fork and try writing two or more copies of your text simultaneously, and the likelihood is that you will soon abandon the project altogether as being all but impossible.

Such a device was hardly likely to rival the musical typewriters invented by Creed and Unger, however primitive one might consider even those to have been—for typewriters they undoubtedly were. At the time I wrote *The Writing Machine* I rather tended toward the view that a typewriter, as I then understood it, ought to print letters of the alphabet or symbols rather than merely longer or shorter lines. On reflection, however, there is no logic to this view and in any case later generations of stenographic machines produced messages of dots and dashes without in any way compromising their inclusion in our history.

We have to wait until the beginning of the 19th century for the first evidence of a typewriter in its more conventional sense, which gives one an idea of just how far Creed and Unger were ahead of their time. But the rest of the 18th century was not without interest by any means, for it saw the emergence of a number of truly remarkable devices which were, once again, not strictly writing machines but were mechanically related to them, and were destined to make a major impact on all who saw them.

The reference of course is to the spectacular creations of Jaquet-Droz and Henri Maillardet. Theirs were all instruments of great beauty and complexity which could be programmed to print a given text, a few words or lines in length, so they do not strictly fulfil the requirement of the typewriter in that they did not produce text at the will of the operator a letter (or group of letters) at a time. They were automata, in fact—of large proportions as made by von Knauss or miniaturized and fitted inside the body of a doll as in the case of the Jaquet-Droz and Maillardet instruments. Operated by clockwork, they featured complex stacks of cams and followers controlling the movements of all the different parts, including those that wrote the text. Technically and aesthetically, these mechanical masterpieces humble any writing machine ever designed or made, but of course they are not strictly in the same category and are related more to the computer than the typewriter by virtue of their need to be programmed.

Friedrich von Knauss appears to have been the first inventor of such a device in 1753 and we are fortunate that several of his fascinating machines have survived. One is in that very special collection of historical instruments in the Istituto e Museo di Storia della Scienza in Florence **(27, 28)**; another is in the Technical Museum in Vienna **(29)**.

27. von Knauss (Courtesy of Istituto e Museo di Storia della Scienza, Florence, Italy)

Von Knauss was a scientist and technician by training and inspiration—twenty years of work went into the creation of his first writing automaton in which a hand, protruding from a huge metal casing containing the movement, first dipped a quill into an ink well and then slowly and laboriously printed a five-word message on a sheet of paper, to the utter amazement and incredulity of all who beheld it. Five years later, a second and more refined model wrote a longer message in a shorter time, while a smaller third machine, the Florence model with its distinctive repoussé facade revealing only the hand and quill, refined the operation even further. Completed in 1767, it was presented by the German Kaiser to the Grand Duke of Tuscany and it has been in Florence ever since.

A plaque at the rear of this automaton is inscribed 'Fridericus de Knaus invenit et fecit,' which may help explain why his name is sometimes written with a single terminal *s*.

But the best was yet to come. The fourth and last von Knauss machine was presented at the Austrian court in 1760 and delighted all who witnessed it by composing the following complex and impressive message:

28. von Knauss (Courtesy of Istituto e Museo di Storia della Scienza, Florence, Italy)

29. von Knauss[20]

Monsieur, Faites-moi la grâce de m'écouter à ce que je vous écris par celle-ci. Le monde a cru, que je ne serais jamais perfectionné par mon créateur, même on le persécuta tant, qu'il fut possible: mais maintenant il m'a mis dans un tel état, que j'écris toutes les langues, malgré tous ses envieux, et je suis en vérité, Monsieur, le plus fidel Secrétaire. Vienne le 4 octobr. [20]

(*Author's Note:* Quoted foreign passages in this book are provided, not for the significance of the meaning of the words themselves, but to demonstrate examples of the long and complex text which was, amazingly, produced at this time. However, for the curious, I have provided a translation of such passages.

(Sir, favor me by paying attention to what I write to you hereby. The world believed that I would never be perfected by my creator, however much he was urged to make it possible: but now he has enabled me to write all languages, in spite of all those envious of him, and I am in truth, Sir, the most faithful secretary. Vienna, 4th October.)

The instrument was some six feet tall, with the movement enclosed in a globe three feet in diameter, composed of six segments hinged at the bottom so that the entire movement could be easily exposed. Inside, a stack of cams turned by clockwork transmitted the required movements to the hand of the automaton by a series of followers and levers.

It sounds so very simple, condensed in those few words, that it might be worth relating in a little more detail precisely what was happening within the machine itself.

Thirty-four levers can be distinguished…the end of each pushes on a large horizontal cylinder…this has a ratchet wheel of seventy-nine teeth controlled by a double click…on the circumference of the cylinder there are 34 rows of 79 holes into which pegs can be put as required. Each lever presses on a rocking lever with a key attached to it. The lines of 34 keys are placed above a system of 34 cams arranged in sectors…a peg is put into the required hole on the cylinder…this releases the corresponding lever and allows the key on the rocker to press upon its cam; the arc as it turns, transmits to the automaton's hand the movements necessary to make the letter by a system of rods…all the other rockers are lowered and therefore immobile. Each sector is made up of three segments of about 30 degrees relating to movements in the three directions…The cylinder moves tooth by tooth; as it goes through one seventy-ninth of its circuit…the table moves one step while the letters are written during the halts. The change-over from one line to the next is likewise automatic, and at regular intervals the pen is dipped into the ink… [17]

As if this complexity were not enough, the inventor was able further to astound his royal patrons by re-programming the device, so that 11 days after its debut it was writing the following totally different but no less immodest message:

Monsieur, tant parfait, que je fut il y a vingt ans, suis-je même encore à cette heure par le grand génie de mon inventeur, et toujours prêt à tous présents, d'écrire tout ce qu'ils puissent désirer. Glorifions donc la Providence de corps et d'esprit, pour qu'elle bénisse le bon dessein de mon créateur, d'oser faire publier au grand monde, mon art, et mes utiles services, par une description générale imprimée, afin que l'on sache, que je sois le premier parfait au monde de cette nature, et que je sois encore, malgré tous ses envieux, toujours Monsieur le même plus fidel Secrétaire. Vienne, le 15 Octobr. [20]

(Sir, I am still as perfect even now as I was twenty years ago, through the great genius of my inventor, and always ready for all those present to write everything that they desire. Let us therefore glorify Providence in body and spirit that it may bless the good intentions of my creator in daring to proclaim my art and my services to the whole world and that I am still, in spite of all the envious [ones], always Sir the same most faithful secretary. Vienna, 15th October.)

'…*le grand génie de mon inventeur…le premier parfait au monde de cette nature…*' No mean boasts, but nevertheless truthful enough and important for us to remember, in the overall context of the historical period we are discussing. We very often read that the later nineteenth century inventors of typewriters of even modest designs and pretensions were simply overwhelmed by the alleged complexities of their task. Yet there was von Knauss already producing his wonderful machines more than a hundred years earlier, in the second half of the eighteenth century. One might well be forgiven for venturing to state that the later inventors were simply lacking in the skills and determination necessary to cope with the mechanical problems they faced, of which, considerably more will be described later.

It is hardly surprising, then, that von Knauss caused the stir he did and that the ink on the automated messages had barely dried before others were demonstrating that the skills required were by no means unique to one man alone. Other automata were sighted almost immediately (a man called Payen is known to have made one), but all were soon to be eclipsed by the creations of the most famous of them all, those of the celebrated Swiss clock maker Pierre Jaquet-Droz.

Born in 1721, Jaquet-Droz was already at the peak of his career when he took the von Knauss idea a step further in 1772 by miniaturizing it and locating it within the body of a doll a mere 30 inches tall and capable of writing (*inter alia*) texts such as the one in the accompanying figure **(30)**. The result was spectacular in more ways than one, for not only was the reduced size of the instrument a mechanical feat in its own right but it also made it easily and readily transportable. And transported it was, throughout the length and breadth of Europe from Russia to Spain and even across the Atlantic to the United States.

30. Jaquet Droz automaton writing sample (Courtesy of Musée d'Art et d'Histoire, Neuchâtel, Switzerland)

Souvenir des Jaquet Droz neuchatel.

Furthermore, Jaquet-Droz disseminated his genius among an important group of collaborators and assistants which included not only his son Henri Louis, but others such as Henri Maillardet and Jean Frédéric Leschot who were talented clock makers in their own right. Not all of their creations were writing automata, but a few were and some examples of these have survived. One is in the Peking Museum, and I would have liked very much to reproduce a modern photograph of this device which is little known in the West. Regrettably, however, my request to the authorities of that Museum for a photograph resulted in a demand for payment of a fee of such monumental capitalistic magnitude that I thought they must have misunderstood my request: what interested me (I wrote to them in reply) was merely a photograph, not the purchase price of the object itself (which possibly puts paid to any personal plans for visits to China).

As it is, the device which at one time formed part of the old Palace Collection is described and illustrated in Simon Harcourt-Smith's 1933 catalogue and the accompanying illustrations (31, 32) are from that source.[40] The writing automaton is located in the base of the clock between the four pillars; resting on one knee beside a small table, it uses a brush dipped in ink to write eight elegant Chinese characters (33) which allegedly translate as *'The eight quarters of the world look hithward* (sic) *and reform; from the nine continents kings approach,'* which no doubt has lost more than a little meaning in the translation.

32.   Peking Museum clock[40]

八方向化九士來王

33.   Peking Automaton text[40]

31.   Peking Museum clock[40]

Now, the interesting thing is that the signature on the movement of the above instrument is that of the London clock maker Timothy Williamson who is known to have made a number of clocks and watches for the Chinese emperor in the latter half of the 18th century. The strong suspicion exists, therefore, that the writing automaton was added at a later date. There are precedents for this sort of conversion in other related fields, the predilection of Chinese emperors for all kinds of spectacular automata being well documented. It is therefore more than likely that the automaton itself was a second generation device inspired by Jaquet-Droz, and it is to be hoped that confirmation of this hypothesis will one day be forthcoming when the article is competently examined.

Another beauty from the same period went on a different pilgrimage altogether, and vanished from sight for 100 years before resurfacing in Philadelphia **(34)**. A resident of that city told the then director of the Franklin Institute that it belonged to his family and described it in some detail. Before the museum could purchase it, however, the man's house burnt down and the automaton was damaged almost beyond salvage, or so it seemed. Nevertheless the Institute bought the remnants and proceeded to restore the movement to working order; no sooner was this completed than it wrote a short four-line poem followed by the message '*Ecrit par l'Automate de Maillardet.*'

34. Maillardet automaton (Courtesy of Franklin Institute, Philadelphia)

36. Fantoni-Turri (Courtesy of State Archives, Reggio Emilia, Italy)

35. Fantoni-Turri (Courtesy of State Archives, Reggio Emilia, Italy)

Yet another fascinating 18th century development, which spilled over into the 19th century, was the emergence of the many mechanical fakes and frauds perpetrated by skilled artists on the credulous public of the day. The writing automata of von Knauss, Maillardet, Jaquet-Droz, and others may not have been strictly writing machines as we now define them but their mechanical value is surely beyond dispute. With a ready market for such wonders begging to be filled, however, it is hardly surprising that a healthy crop of fakes was very soon in evidence.

Wolfgang von Kempelen was one of those who obliged in this respect, with 'automata' which displayed a baffling montage of wheels and pulleys fitted to a large cabinet containing several doors. These were opened one after the other to prove that the cabinet was empty; meanwhile, the operator inside was able to conceal himself by passing from one compartment to another. A magnet, prominently displayed on the outside of the cabinet, suggested an explanation for the gullible—magnetism, after all, was very much in vogue at the time. Von Kempelen was only one of many to cash in on this market and everyone had his own particular angle. Even one of Jaquet-Droz' automata, which had passed through the hands of Maillardet, finished up in the possession of no less a character than the famous Robert-Houdin who had it, or rather his concealed accomplices manipulating it, writing or drawing anything requested by the audience.

A little of everything then appeared in the 18th century—everything from frauds and fakes to the very first writing machine to qualify as such. Significantly this writing machine was designed to reproduce musical scores—significantly, because it was not until well into the 19th century that society was ready to accept anything that was not written by hand, so much so that it was even considered an insult to send someone a typewritten letter until relatively modern times. This lingering prejudice was not altogether a bad thing at first, because it was a pain in the neck trying to extract a legible piece of text from an early typewriter. Spacing was erratic, keys jammed, paper creased and crinkled up, and methods of inking required varying degrees of luck and skill so that the result was generally all but illegible copy. It is worth remembering that you simply could not buy what you wanted from the local store, in those days. If you wanted a duplicate of the original, it was necessary to make 'black paper,' as carbon paper was often called, for your writing machine. Unless you lived in London, or otherwise had access to some of Messrs. Wedgwood's products, this had to be done at home by mixing fat and lamp black, rubbing it onto a sheet of paper, and laying it aside for a long time to dry out. If a ribbon was required, a strip of material was treated in a like manner. Ink was plentiful enough, but while it was fine for quill pens it was not exactly right for mechanical applications. The available paper was excellent for manual use, too, but far from ideal for a typewriter, as early surviving typewritten correspondence such as the Fantoni-Turri letters (35, 36) so graphically illustrates.

More on Turri and Fantoni in due course. Meanwhile, a short pause to discuss carbon paper and typewriter ribbons, since clearly these two articles will henceforth appear so frequently throughout the following pages. In the days before retailers supplied them, ribbons had to be made by soaking strips of cloth in some kind of ink-based solution and hanging them up to dry. Simple enough, except that to be effective, a typewriter ribbon needs to be of a certain length, even in modern times with modern materials and manufacturing processes, otherwise the ribbon will need constant replacing. In early times, everyone had to make their own typewriter ribbons and the seemingly endless strip of ink-soaked cloth had to be stretched and snaked and draped around a room, over and under furniture and fittings, and left *in situ* for days or even weeks at a time in order to dry out, before being wound manually onto spools or bobbins for use in the machine.

37.   Wedgwood leaflet

R. WEDGWOOD's PATENT
Manifold-Writers and Penna-Polygraphs,
MADE AND SOLD FOR MONEY ONLY
At 328, Oxford Street, and at his Office, 17, Tokenhouse Yard.

Enumeration and Description of Inventions founded on the above Patents.

MANIFOLD WRITERS produce several Fac-similes, and are made in the Form of Books, Portfolios, &c.

MANU-POLYGRAPHS are adapted for preserving Fac-similes of Writings made on Horseback or in a Carriage, and in using are conveniently held in the HAND without the aid of a Desk or Table.

NOCTO-POLYGRAPHS enable the Blind or others in the Dark, by one Effort, to write Single, Double, or Treble.

PEN AND STYLO-POLYGRAPHS produce several Fac-similes of Writings, by an Instrument uniting the Pen and Style, called the Fountain Style or Penna-Triplex.

VIA-POLYGRAPHS—By these several Fac-similes of Writings or Drawings of any Size may be made in the open Air.

NOCTO-VIA-POLYGRAPHS are a convenient combination of the Nocto & Via-Polygraphs, & Manifold Writer.

BOOK PENNA-POLYGRAPHS produce Duplicates of Writings with Ink, on a prepared Book, & common Paper.

SHEET PENNA-POLYGRAPHS produce Duplicates of Writings with common Ink on Sheets of common Paper.

ROLL PENNA-POLYGRAPHS produce Duplicates of Writings in common Ink on Rolls of Parchment; also several Copies or Fac-similes of large Charts, &c.—Large Charts, &c. may also be copied by them.

POLYGRAPHIC DESKS AND TRAVELLING COMPENDIUMS contain various Inventions for increasing Writings or Drawings, which are in the most portable Form, together with a complete Set of Dressing Instruments.

PROTEAN TYPE, or STEREOTYPE TABLETS—By which a Person forms by a short and secret Hand-writing, in a Language founded on Philosophical Principles, a STEREOTYPE from which Copies may be multiplied *ad infinitum*

PROSPECTUS OF AN EXPLANATORY TREATISE

ON THE

PROTEAN TYPE, OR STEREOTYPE TABLET,

By R. WEDGWOOD, the Inventor, 328, *Oxford Street, and 17, Tokenhouse Yard:*

SHEWING—I. An Art of composing in the act of Writing, by a Fount of Types, which may go in the Pocket. II. An Art of Writing, which may be attained perfectly in a Moment, and which admits of inviolable Secresy. III. A Writing and Printing Character, the shortest yet invented, perfectly uni-vocal, and easy to be formed. IV. A Rule of Language philosophically just, and free from exceptions.——In this Work is also noticed, a System of Education tending to create perfect uniformity of Idea, and to give security against the false notions, ignorance, or prejudices of the Teacher, whilst it excites agreeable sensations and close attention in the Pupil.

CONDITIONS.—This Work will be printed in Folio, on a fine Wove Paper, bound in extra Boards, price One Guinea and a Half, to be paid on Delivery.—As soon as 300 Copies are subscribed for, the Work will be put to Press.—The Subscribers Names will be received at No. 328, Oxford Street, and at No. 17, Tokenhouse Yard; and it is particularly requested that those scientific Persons who intend to promote the above Invention, will be pleased to cause their Christian and Surname, with their Place of Abode, to be signified, in order that every Copy may be delivered in the order subscribed for, as well as that a List may be accurately printed to precede the Work.

W. M. Thiselton, Printer, Goodge Street, London.

Carbon paper...or *carbonated paper*, as it was originally called...had a somewhat different history in that, thanks to Mr. Ralph Wedgwood, it was 'made and sold for money only' *(sic)* in his premises at 328 Oxford St. and 17 Tokenhouse Yard, London **(37)**. Wedgwood was not only the inventor of carbon paper—the watermark on an original sheet of his *carbonated paper* in the author's possession shows the name of the paper maker and the date **(38)**—but also of a considerable number of copiers, gadgets, and appliances which used it. His patent, No. 2972 of 1806—granted (no doubt on one of his better days) by 'His most Excellent Majesty King George the Third'—was for 'Producing Duplicates of Writing.' The patent specification states

> ...In writing by this mode, I make use of a prepared paper, which I call duplicate paper. This is made by thinly smearing over any kind of thin paper with any kind of oil, prefering those kinds of oil which are least liable to oxygenizement, or to be evaporated by heat. The ink made use of in this mode of writing consists of carbon, or any other colouring substance, and finely levigated in any kind of oil. This ink is to be evenly spread on leaves of thin paper, or any other thin substance, after which it should remain for five or six weeks, or any shorter period, betwixt sheets of absorbent or blotting paper, after which it is fit for use. This I call carbonated paper...To write singly in this mode, I lay a leaf of the carbonated paper upon a smooth tablet of metal, or any other smooth substance, and over that I lay a leaf of duplicate paper; upon these I write with the style,' (i.e. stylus) 'the pressure used causing a transfer of the carbon to be made upon the under side of the duplicate paper, which being transparent instantly appears through as if written upon the paper. To write double, I lay, first a leaf of writing paper upon the tablet; second, a leaf of carbonated paper upon that; third, upon both a leaf of duplicate paper; and upon the papers so disposed I write with the style. The effect produced is a double transfer of the carbon from the carbonated paper, that on the lower surface thereof to the letter paper beneath it, and that from the upper surface thereof, to the under surface of the duplicate paper above it...

Perfectly clear, simple and succinctly phrased...particularly that final sentence! But then, clarity of meaning was not something dear to the hearts of early patent clerks, attorneys or agents so it is hardly surprising that we find examples of sentences in Wedgwood's patent typically consisting of 240-odd words from one period to the next, the sentences further adorned with lush verbiage such as '...*anything therein-before contained, to the contrary thereof in anywise not withstanding...*

Back, then, to the writing machines themselves: the one factor which mattered little, if at all, to early typewriter inventors was operating speed. Which was just as well, because early typewriters were slow...very slow! But then, speed in general was not an issue in those times. Writing itself is a slow process, even today when we have come to accept illegible scrawl as a fact of life. In past centuries, professional scribes prided themselves upon the beauty of their script and dictated the standards for all to match. The earliest typewriters would probably have been capable of speeds more comparable to contemporary handwriting than to modern typewriting, due in no small part to the need for constant mechanical attention and adjustment which the early machines certainly required.

But music, as we have said, was a different matter. It is interesting that the first writing machine should have been one designed to type music rather than text, because of the immediate applications for such a device. Theoretically speaking, anyone wishing to compose or copy an existing musical score had only to play it on the keyboard of one such writing machine and it was instantly on paper, without the need for constant stopping and starting. The manual dexterity was there, and of course pianists were capable of playing notes far quicker than early machines were capable of recording them—a factor not lost on later inventors, many of whom designed conventional typewriters with keyboards taken straight off their parlour instrument. In fact a number of these inventions looked much more like spinets than typewriters: Francis **(39)** and Michela **(13, 14)** spring immediately to mind.

However, although operating speed did not matter at first, there was one application in which it did and this led to the development of a different breed of machine altogether. Typewriting, which required many keys and moving parts, was doomed to remain a slow and uncertain process. Tackling this problem necessitated not only improvements in design but also manufacture to tolerances behind the scope and ability of early inventors, most of whom had little mechanical or technological expertise. From the very earliest days of industrialization, however, a machine was needed capable of directly committing the spoken word to paper. If business and commerce did not immediately realise the potential of such an instrument, it was certainly not lost on those responsible for recording legal proceedings or the speeches of politicians at local and national gatherings and of delegates at trade and pro-

38. Wedgwood watermark

39. Francis (Courtesy of Smithsonian Institution)

the simultaneous depression of two or more keys, and the advantages of such a method are immediately obvious.

This leads us naturally to perhaps the most interesting and important application of these principles in the infant writing machine, namely as an aid to the blind, which in turn brings us to the first machine with typed text and not musical scores. The importance of a writing machine for people who are sightless is obvious, for without it they are to all intents and purposes unable to correspond unless they use a sighted person as a scribe. Writing frames of different designs had been known for centuries but their use was limited and the results haphazard. Once perfected, typewriters for the blind—particularly those which embossed, permitting a blind person not only to write but also to read the text—became universally available. At the same time shorthand machines with their condensed keyboards and endless rolls of paper opened up important employment opportunities for the blind in the field of stenography, sight being no advantage whatever to the operator of such devices.

The first machine for the blind, indeed the first typewriter *ever* made which typed letters rather than musical scores, was the invention of the Italian Pellegrino Turri and the scenario would make a perfect libretto for an Italian opera. Turri was born in 1765 in Reggio Emilia. His family was well known and his life is adequately documented. Married, unhappily one suspects, to a wealthy and well connected woman, he entered public life with moderate success which continued until his death in 1828. Meanwhile he had met and taken more than a passing fancy to the young Countess Carolina Fantoni who was some sixteen years his junior and by all accounts very beautiful. She was, however, completely blind. Contact between them was not difficult while she remained in Reggio, but she left in 1808. Her departure was obviously not unexpected, for by then Turri had devised and built for her what must have been a quite superlative writing machine, given the quality of work produced on it by a blind woman whose hands were already crippled by arthritis.

We can only speculate as to the design of the machine, for it has not survived. But the work it produced has, and a considerable body of it was preserved in the State Archives in Reggio Emilia. I say 'was' advisedly and I assume that at least part of this file is still there. When I first discovered its existence more than twenty years ago and publicized its importance in my first book, it contained correspondence which has since vanished. The file at that time contained thirty-one typewritten sheets—a considerable body of work which included not only letters but also original compositions and poems, although from references in the correspondence itself, it was already clear that it was far from complete even then. In fact an original letter, reprinted in an Italian journal in 1908 to mark the centenary of the machine, was itself missing from the file when I first opened it.[56] The author of the article in which the letter was reprinted was none other than the then director of the Archives, Umberto Dallari. Presumably the letter was removed at the time the article was written nearly eighty-odd years ago and never replaced, its historic importance perhaps not fully appreciated. This excuse can hardly be offered for more recent events, however, and the file will continue to suffer depletions unless adequate safeguards are introduced.

Now, the historic importance of this body of early typewritten material can hardly be overstated. Not only is it the first evidence of its kind, but it is of a quality and character which leaves one quite breathless with amazement. The sheer excitement I felt nearly a quarter of a century ago when I first opened the file and examined the contents is an experience I will never forget. I was already reasonably knowledgeable by then, and familiar with the typewritten products of many early machines, both prototypes and production models. I was also aware of the many problems which

fessional conferences and congresses, all of which seemed to be increasing in number every year. No conventional typewriter could hope to manage it—not even today—but one which dispensed with orthography altogether had a chance…and stenography was born, almost overnight, as it were.

As might be expected where there were no precedents and new ground was being broken at every step, the diversity and variety of ideas and applications is almost too extensive to record in detail. Virtually every inventor devised his own method and alphabet and its success or failure was not always dictated by what, in retrospect, appears logical or reasonable. Signs and symbols ranged from a series of dots to complex geometric patterns, many using letters of the alphabet in often baffling combinations. The machines themselves ranged all the way from tiny hand-held objects to the inevitable piano keyboard instruments.

The one feature that most of them shared was the chord principle by which characters, or groups of characters, were printed by early inventors had faced: inking methods, spacing, alignment, paper feed…to name but a few. Many of these late 19th century inventions were so well documented in the notes and letters of the inventors and their families, assistants and financial backers, that we have a graphic record of the many agonies and few ecstasies of a typewriter inventor's existence. (Sholes springs immediately to mind, as does Ravizza…and many others. Sholes at least lived to see his efforts crowned with success; Ravizza was to be denied even that ultimate compensation.)

Yet more than half a century earlier Turri, starting from scratch, was able within the space of a year or so (as we can deduce from surviving information) to design and build a typewriter for a blind woman with crippled hands, producing work of such high quality and on paper utterly unsuitable for the purpose, that all superlatives are quite inadequate!

What exactly has the Countess Carolina left us? The thirty-one typewritten sheets which I was able to examine back in the 1960s revealed a number of fascinating details as is evident in the accompanying illustrations **(35, 36)**. We can see immediately that Turri had problems even without reading the text of the letters, for some of the pages were typed in ink while some were not. Turri had originally designed the machine to use 'black paper' (carbon paper), of the kind already invented by Wedgwood in 1806. Since it was clearly not possible to buy this commercially in Italy at the time, it had to be made and, it may be safely assumed, the inventor himself had to make it. A tiresome process for, unlike modern carbon paper, a single impression would have removed all the carbon from a given spot and a regular supply would have been needed to keep the Countess happy, and indeed the surviving correspondence bears eloquent witness to this fact.

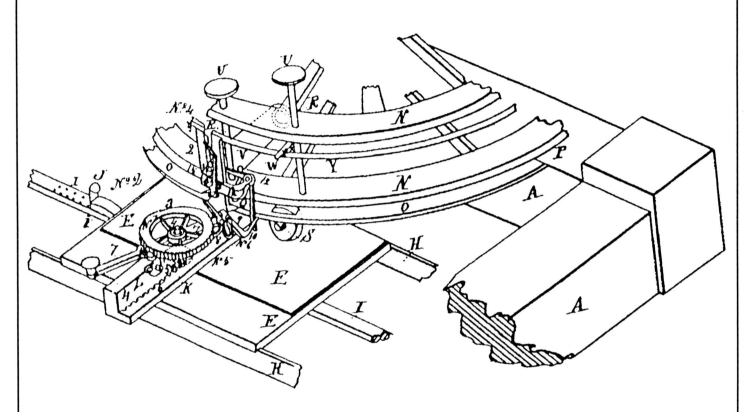

40. Thurber, 1843 patent (detail) (Courtesy of US Patent)

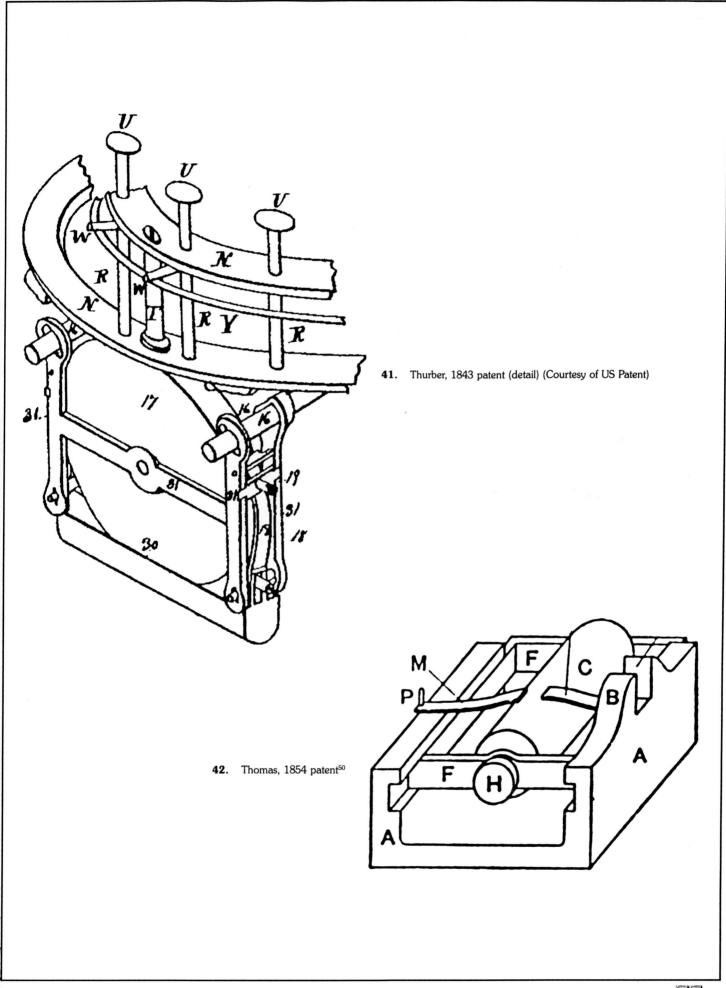

41. Thurber, 1843 patent (detail) (Courtesy of US Patent)

42. Thomas, 1854 patent[50]

But then, within the same archives we find a later letter with a very different appearance altogether. It was the first attempt at using the so-called 'new characters' which Turri provided (and presumably fitted) because the sharp letters needed for carbon paper were found to be unsuccessful with ink which required flatter surfaces. However the results produced by the modified type face, if anything, were worse than before, for the thin paper sucked up the ink like a blotter and the frugal Countess infuriatingly insisted on typing on both sides of the page.

Furthermore, whatever new system Turri devised was more or less good for one page only and often became fainter even as the page progressed.

But we are being excessively critical. The instrument was far from perfect, but under the circumstances it was not just good, it was great! One wonders what might have transpired had the whole thing not been abandoned, for the correspondence ceases abruptly in 1810, some months after Carolina announced her intention to marry, and the fact that the correspondence ended immediately after her marriage leads one to deduce that maybe her husband was responsible, suspecting perhaps that the relationship between Turri and Carolina had not been altogether platonic.

That is the story of Pellegrino Turri, then, and the first ever machine that typed letters rather than musical notes as well as the first typewritten correspondence and the first typewritten manuscript. When Turri died in 1828 he left the Countess's letters to his son Giuseppe who, on his own death in 1879, in turn left them to the Archives of Reggio Emilia. Carolina kept the typewriter until she died in 1841, whereupon it was returned to Giuseppe and subsequently lost without trace.

Maybe, some day, someone will rummage through the attic of an old house in Reggio Emilia and stumble across it. Stranger things have been known to happen!

Chapter Two—
Second Generation:
The Age of Manufacture

So there we have it—the few significant events which spanned a century, from Mill through Creed through Jaquet-Droz and his associates, to Turri. Not much to show for a hundred years of development, but we have been dealing with the 18th and not the 19th century and significant changes were just around the corner.

What began at a trot soon broke into a gallop, with a proliferation of inventions and ideas almost too numerous to record. There is little to be gained from a detailed analysis of each and every development, some of scant intrinsic value—for the sake of historical record only, these are listed in Chapter Five. Others, however, require more particular attention, because machines were now beginning to take on distinctly recognisable typewriter characteristics, as inventors in different parts of the world grappled with common problems and solved them—or tried to—in one or more of a number of well-defined means.

Within a mere forty or so years after Turri's historic achievement, a typewriter production line was ready to mass produce

machines in the United States. One was actually doing just that in England, on a small scale, and another in France.

By that time, every major design element in manual typewriter history had already been invented in one or more industrial nations across the globe. Keyboards, type bars, type wheels, daisywheels, carriages, platens, feed rollers, upper and lower case, ribbons, space bars, proportional spacing, stenographic machines, machines for the blind, telegraphic typewriters…all had been invented, time and time again. By and large it is fair to say that after 1850 or so, there were no new mechanical surprises; by and large, what innovations there were relate more to the inventors and their struggles rather than to the objects they had designed. The surprises, in other words, were human rather than technological.

By 1854 the first electric typewriter was invented, and that was it as far as major innovation was concerned, until the introduction of word processing. And in fact even elements in word processing hark back interestingly to the past. Storage of typewrit-

43. Hansen's Skrivekugle (Courtesy of Sotheby's, London)

ten material for later use, for instance, is directly related to the early 19th century telegraphic designs in which messages were first stored on perforated paper tape and later fed into the printing machine as and when required—the system proving so efficient that it was used almost until the present day.

England, the United States, and France were the three countries hosting typewriter manufacture virtually simultaneously around the middle of the last century. Given the world-wide interest in the development of writing machines by inventors at least, if not by the public at large, one might assume that the machines from these three countries would share common design features and characteristics, all tried and tested in the course of the preceding decades. There were after all more than fifty separate inventions from the days of Pellegrino Turri and Carolina Fantoni and many of them were patented and publicized. Cross pollination between inventions was commonplace enough even in those days to cause massive national and international disputes which reverberated well into the 20th century.

And yet, the three mid-19th century designs which actually reached the stage of manufacture could hardly have been more different.

The Frenchman Pierre Foucauld is worth first mention, not only because of chronological priority but because he appears to have been the most dedicated. He began his first design as far back as 1839, or rather he launched his first design in that year, for exactly when he started on it is not recorded. As with so many of his contemporaries, it was the plight of the blind which provided the inspiration. He was himself blind, and corresponding with those who could see was his main preoccupation. The product of his efforts was a device of what we now call 'radial plunger' conception (16): this, along with other terminology used to describe typewriter design principles was first introduced in *The Writing Machine* and these designations have become universally adopted ever since. I do not propose to repeat the detailed analysis of this subject which was made at the time, and students who require further information might well be advised to refer to the source.[1]

In the plunger principle, the character to be printed was at the end of a rod or bar which was thrust downwards directly onto the paper either by a finger or by a selector, without the intermediate intervention of any kind of linkage. In radial plunger machines, these rods were located in such a way that they fanned radially outwards from the printing point in some form of framework which supported them vertically. This was not a particularly popular idea in typewriter history, although some exceedingly fine machines were made from this design which reached its pinnacle in the superlative Hansen Writing Ball—the Skrivekugle—of the 1860s, manufactured in some numbers well into the final quarter of the century (frontispiece, 4 et al.).

Foucauld's first device was an interesting combination of principles, all of which had been tried before—if you analyse the components, not a single one was original. This is hardly surprising, as we have already noted that all major elements had been invented at a very early stage indeed. Foucauld's achievement lies in the manner in which, despite his blindness, he assembled the various elements. Calling his invention Raphigraphe (16), he positioned ten plungers in an arc of about 45 degrees. Each of these plungers ended in a point which embossed the paper in a different spot; by pressing several plungers at the same time (and back-spacing for additional marks when required) Foucauld was able to build up the shapes of all the letters in the alphabet.

Everyone who saw this machine, including Braille himself apparently, was happy with it, but not so the inventor. It seemed to work well enough for the sighted but not for the blind operator who had trouble reading back what he had written, and this was a serious disadvantage not only with long communications or with those which could not be completed at a single sitting, but also with those requiring corrections to the text. The trouble was that the letters embossed by the machine were too small for a blind operator to feel. They simply had to be increased in size to be worth anything at all, but if they were, then postage on an ordinary letter might prove quite prohibitive, especially as it is obvious that only one side of a sheet of paper can be used with embossed text.

The answer, or so it seemed to Foucauld, was to type two originals at the same time, so he took one Raphigraphe and placed it upside down on top of another, with a single set of keys simultaneously operating both instruments. The lower of the two, fitted with small blunt-ended plungers, typed a visible message for the sighted, at the same time as the upper one embossed larger letters for the blind.

This was not much good either, and once again the inventor went back to the drawing board, managing to put past experience to good use in his third effort in 1849 which he christened Clavier Imprimeur. The radial plungers were retained, sixty of them in two parallel rows, with a space bar in between. The plungers now typed actual letters, not dots, and there were lots of small refinements such as the all-essential bell to indicate the end of a line.

The first reports of this invention were published in 1850 and a prototype won a gold medal at the Great Exhibition in 1851. Limited production began soon afterwards.

Meanwhile the Englishman G. A. Hughes, the second of the inventors to share the honours of manufacture, had completed a simple and absolutely beautiful instrument which was also launched at the 1851 Great Exhibition in London. Hughes was Governor of the Manchester Blind Asylum and his invention was primarily designed for typing embossed letters, although carbon paper was used to produce a visible image. Once again, plungers were used but, unlike Foucauld, he positioned these vertically in circle around the periphery of a circular housing, with raised letters around the outside which were easy for a blind operator to select. A lever, prominently positioned, pressed the plunger onto the paper beneath and the whole unit travelled along a horizontal rod one space at a time. A thumb screw centrally positioned on the base at the front of the machine was given one complete turn at the end of each line, thereby advancing the paper the desired amount.

The Typograph (2, 44), as Hughes called his invention, also won a medal at the Great Exhibition. Manufacture soon began and was still continuing over ten years later when the machine was again displayed, virtually unchanged, at the 1862 London Exhibition. Just how many were manufactured has never been fully established but several are known: there is an example in the London Science Museum collection, and the one in the accompanying illustration bears the serial number 52, so production was presumably somewhat larger than the rarity of surviving examples would appear to indicate.

Hughes ultimately enjoyed more success with his product than did Foucauld with his, and this is probably as it should be. On the other hand, the third inventor to embark on manufacture at the same time was the American John Jones and he, regrettably, enjoyed no success at all. He was in the process of completing the first 130 examples of his invention when the factory went up in flames and all the machines were destroyed, a prototype in the inventor's home and the patent model being the sole survivors.

What Jones was in the final stages of launching onto the market in 1852 was a Mechanical Typographer consisting of a cylindrical platen above which was a type wheel, circular index, and indicator arm. The platen moved in a linear direction as typing progressed, just as it does on modern machines. While this arrangement was by no means original to Jones, neither was it as obvious at the time as it would seem to appear today. Decades later people were still designing typewriters which printed around the cylinder rather than along its length. Yet another praiseworthy though not unique feature on Jones' invention was the provision for differential spacing.

Jones persisted a little longer, the fire in his heart not extinguished with the one in his factory, but the circumstances were no more propitious the second time around, and manufacture was not tried again. Perhaps it is just as well, for on this later Domestic Writing Machine, which he patented in 1856, he abandoned the cylindrical platen in favour of an endless belt. Future generations will doubtless remember him more for his first than for his second patent.

From the one-off Turri machine to the first typewriter production line a mere forty or so years had elapsed, but they were action-packed years as far as typewriter history and development were concerned. No fewer than fifty names were added to the rolls during that period—many hundreds of machines of every conceivable design provided a massive depository of knowledge from which later inventors were able to draw inspiration and experience. Europe and the United States shared the honours during this developmental period before the balance gradually swung in favour of the latter as inventors on both sides of the Atlantic grappled with common problems to which they found vastly differing solutions.

As the century progressed, the commercial pressure on inventors to protect their ideas increased, and patent applications were submitted often before the ink on a rough sketch had time to dry. In fact, commercial considerations began to assume paramount importance: if you had a patent, you were in business, otherwise

44. Hughes' Typograph (Courtesy of J. G. Foster Collection)

you had nothing. Backers with good money to spend became increasingly easy to find; companies were born, investments made, shares sold, extravagant claims and promises received wide publicity. Conti, Burt **(55)**, Galli, Drais, Thurber **(40, 41)**, Ravizza **(15)**, Michela **(13, 14)**, Mitterhofer **(24)**, Hansen **(58, et al.)**— to mention but a few—each made his greater or lesser contribution, accelerating the pace of invention towards mass production in the last quarter of the century. Each one added something to the pool of knowledge from which subsequent inventors drew.

What were these individual contributions? Keyboards provide one example. One might expect to find a high degree of unanimity among early inventors in designing keyboards because, as we have already mentioned, the dexterity of a musician manipulating a keyboard was plain for all to see. It is hardly surprising therefore that the piano keyboard was chosen time and time again. Creed **(25)** and Unger used it, as did the few 18th century inventors who followed, as well as all inventors of musical typewriters for the next hundred years. A Frenchman called Gonod built it into a shorthand typewriter in 1827 and a French piano manufacturer called Henri Pape incorporated it into his invention in 1844. Jacob Brett in England applied it to printing telegraphs as did the great Charles Wheatstone and, a few years later in the States, D. E. Hughes added respectability to the principle by actually entering into production with it in considerable numbers on his famous telegraphic instruments **(45, 46)**. The list goes on and on: Francis in 1857 **(39)**, Guillemot in France two years later, Michela in 1862 **(13, 14)**, then Benton Halstead and John Pratt…Ravizza used it **(15)**…even Sholes tried it…

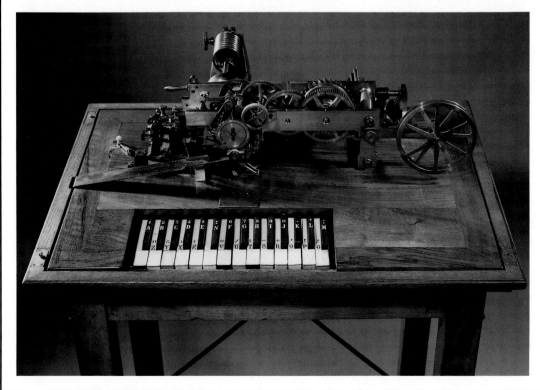

45. Hughes printing telegraph (Courtesy of C. N. A. M., Paris)

But if adapting a piano keyboard were all that easy, everyone would have jumped on the bandwagon and we should never have had anything else. The truth is that it was far from easy. Adapting the piano keyboard to a successful machine to type text presented enormous mechanical problems which not everyone could successfully grapple with. Creed could not, and did not even try; he simply ran paper under a keyboard, more or less. This worked for music after a fashion, but would have been very poor indeed for text. Some people did try to invent machines like that in later years and a number of stenographic typewriters resulted.

But really, this idea was no good for writing text—if your piano keyboard was in alphabetical order and finished with the letter *z*, this would be forever positioned on the far right, even though it might be needed as the first letter of the word. Now, music requires more notes, and therefore more keys, than text requires letters of the alphabet for which a keyboard of a couple of octaves of black and white keys would be sufficient. But even so, each letter had to be made to print at a common point (usually, but not always, at the centre of the machine). In order to achieve this a complex system would be required, even on a reduced keyboard—as witness the underside of the Hughes printing telegraph **(46)**—and levers and linkages and ratchets and so on would be needed to move either the keys or the paper a space at a time for printing to occur in the correct sequence.

The piano keyboard itself was not intrinsically inefficient, of course, even though many inventors could not cope with it and gave up on it altogether. Many preferred keys of small dimensions, similar to those we have today. Beach was one who came up with just such an idea in the middle of the last century, and Pratt some years later switched from piano keys on his early efforts to little cylindrical keys on his last model. The advantages are obvious and we are too familiar with the design in modern times for this to require any further commentary. Ultimately, everyone switched and the modern keyboard was universally adopted. However, far from being the foregone conclusion it might at first appear, we find that piano keyboard machines were still being invented and manufactured in some numbers over the course of quite a few decades in the second half of the last century.

Where the piano keyboard scored heavily was on printing telegraphs using pin barrels, and this was where the concept was most widely applied. Piano keyboards were ideally suited to these machines. The pin barrel was made to rotate in front of the keyboard in such a way that pressing down on a key made contact with its corresponding pin, which stopped the barrel from turning further; a type wheel fitted to the end of the barrel was thereby stopped at the corresponding letter, and printing could proceed. Releasing the key allowed the barrel to continue revolving. The pins on the barrel performed a single revolution in a gradual and regular spiral around the surface of the cylinder from one end to the other so that each key and its corresponding pin stopped the cylinder and hence the type wheel at a given point for the selection of the desired letter.

This was a great idea for a printing telegraph; all that was required was to pass paper tape under the type wheel at the end of the barrel for printing to be performed simply and efficiently. No need for a carriage and its many attendant problems when you

46. Hughes printing telegraph (Courtesy of C. N. A. M., Paris)

47. Mignon (Courtesy of This Olde Office, Cathedral City, California)

print on paper tape; all you need to do is devise a means by which the paper is advanced a space at a time after each printing, which is simple enough. And several inventors not only patented but actually manufactured such machines long before ordinary typewriters had caught up with the technology. Jacob Brett in 1845 and Royal E House the following year were the earliest to be granted such patents. Despite desperate teething problems relating inevitably to the difficulty of synchronizing receivers and transmitters, which was clearly essential, the invention was manufactured and used on both sides of the Atlantic for a number of years during the 1850s and even into the 1860s, until better instruments replaced it.

One of these better instruments was the device invented by D. E. Hughes in 1855 and patented in the United States the following year, which has already been mentioned. This succeeded in a major way, being manufactured in considerable numbers in Britain, France, and Germany and used extensively on telegraphic transmissions for over twenty years **(45, 46)**. That was certainly the greatest success story the piano keyboard ever achieved, although Michela fitted one into his stenographic typewriter **(13, 14)** in 1862 with quite spectacular results and manufacture of this machine continued into the 1880s in both France and Italy. Operating speeds of up to 225 words per minute were reliably reported to have been achieved and this, if nothing else, ought to dispel any suspicions that the piano keyboard principle was a hopelessly anachronistic concept.

Michela's Stenograph, significantly, printed on paper tape. I say significantly, because the difficulty of producing an efficient carriage and escapement was partly responsible for arresting the development of the typewriter for at least a quarter of a century. If

you could print on tape, as in the printing telegraphs and shorthand machines, it was all a great deal easier. Nevertheless the Michela machine had to cope with problems which the telegraph deftly avoided, in that it required direct individual linkage between the keyboard and the keys—no comb and pin-barrel here! The result was a compact instrument in which each hand operated its own piano keyboard of six white and four black keys with the roll of paper tape between the two. The keys were pressed in chords, as required, and printing was by means of a ribbon on the underside of the tape in a shorthand code of symbols. A single roll of paper tape 300 metres long was good for up to seven hours of dictation. Nothing even remotely as good as this was to appear for several more decades.

So there was nothing intrinsically wrong with the piano keyboard, then, and in certain applications it even proved to be an advantage. However, while other inventors were coming up with very different keyboard designs indeed, for some people any kind of keyboard at all was proving too difficult or too expensive to deal with, and in these inventions keyboards were abandoned altogether in favour of a simpler and more direct alternative in the shape of a letter index and indicator arm.

The idea was to have a plate with the letters in whatever order you wished making up a grid of square, rectangular, or elongated shape. An indicator was moved, usually by the right hand, to the required letter on this index which action simultaneously positioned the typeface over the printing point whereupon pressure was exerted, usually by a lever or button operated by the left hand, thereby bringing type and paper together. If this seems like a very slow and cumbersome system, and it was certainly slower than using a key

THE IMPROVED No. 2
COLUMBIA TYPE WRITER.

HIGHEST AWARDS at 1885 Exhibitions at London, New Orleans, & New York.

It is to the Pen, what the Sewing Machine is to the Needle.

THE TYPE WRITER is essentially the writing instrument of the age, its general use is but the matter of a few years, and, Authors, Professional, and Business Men have awakened to the fact that TYPE WRITERS are a necessity, economizing time and giving perfectly distinct writing when at present much time is wasted in trying to decipher Draft copies and letters.

The speed attainable is 40 words per minute more than any other portable machine, thus equaling the best key machines.

We unhesitatingly claim THE IMPROVED No. 2 COLUMBIA to rank first for

Extreme simplicity, 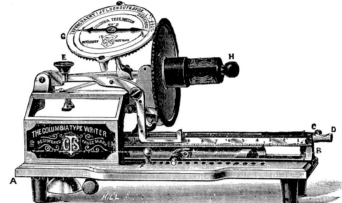 Ease of operation,

Compactness, Durability,

Economy, and Speed.

ONE THIRD ITS ACTUAL SIZE.

A The base of the machine is provided with a groove in which the **B** Paper Carriage runs freely. The rod **C** is lifted up by the knob, and the paper inserted at the back under the roller, turn the knob **D** till the paper is in position for writing, then replace rod **C**, depress spacing key **E** and holding it down, push carriage to left till edge of paper is under type wheel, then place banking pin **F** in the hole on the right of projecting pin on carriage, this ensures a straight margin. **G** is a dial with indicator, on which the letters of the alphabet, numerals, stops, &c are engraved. **H** is a vulcanite handle, which is turned so that the round end of the indicator stands at the Capital letter, or Figure required to be printed, but if a small letter or stop is required, the thin end of the indicator should point to the letter wanted, then depress the handle and the corresponding letter on the Type wheel, will be printed on the paper. **E** the spacer, which, when pressed down, carries the paper carrier **B** back, so that a space is made between the words of an equal distance, when the alarm bell sounds, the end of the line has been reached, so depress the spacer, pull the carriage forward till it reaches the banking pin, then turn the knob **D** one notch, which will throw forward the paper ready for the next line.

The machine is inked by squeezing a little ink from the flask on the Pad, this will last for two or three days. Copies from which can be taken in the Letter Book,

THE IMPROVED No. 2
COLUMBIA TYPE WRITER

Complete in handsome polished wood box, with lock and key, and flask of Ink, weighing under 4½lbs. and measuring 11 x 5 inches. Price **£6**

the best and yet the cheapest upper and lower case TYPE WRITER in the market.

LOCAL AGENTS IN ALL TOWNS, OR TO BE HAD FROM YOUR STATIONERS.

Issued by the Wholesale Agents W. J. RICHARDSON, & Co., Poultry, E.C.

48. Columbia Type Writer leaflet

3 8

board, the gap between the two was not usually as great as may at first appear. Keyboard machines were themselves slow, with heavy actions requiring considerable finger pressure; on the other hand, operators of indicator machines quickly mastered the simple requirements of the system and attained very creditable speeds indeed. Of course, indicators could not compete as a viable alternative to machines with conventional keyboards when these improved in quality and touch typing became the norm, but this did not happen overnight and indicator machines were still being mass produced long after the First World War. In fact a few—the Mignon (47) is the outstanding example—were still being produced up to the Second World War.

Indicator machines, at the time, had a lot going for them. They were on the whole cheaper than their keyboard rivals—some of them were dirt cheap, although others which were more complete and practical tended almost to match their keyboard competitors in price. Moreover, their very design simplicity ensured that they were direct, reliable, and usually free from the bugs which plagued keyboards and type bar designs. The better ones, such as the Columbia (9, 48, 137) with its circular index, were even able to offer differential spacing, a feature which their type bar competitors found great difficulty in matching. Differential spacing is the process, well known in the printing industry, of assigning a wider space for one letter than another, for example an *m* as against an *l*. This feature presented problems of such mechanical complexity to typewriter designers that after the initial flush of enthusiasm for the principle, it was virtually abandoned for the better part of a century until IBM and others 're-invented' it relatively recently.

Indicator machines required the letters of the alphabet to be printed on an index of some form or other. These letters, as indeed the letters on a keyboard, had to be placed in some sort of order…and here every inventor simply let his imagination run riot. Most first generation inventions tended to place letters in alpha-

betical order. This made a certain amount of sense and had its own logic, because the letters were easily found by the novice being introduced to the typewriter for the first time. Since speed was very much of secondary importance—except of course in stenographic inventions—and machines were not mechanically efficient anyway, alphabetical order had a lot to recommend it.

But before long, inventors were spinning away from alphabetical order with wild abandon, as though the very perversity of the letter order on their inventions were a mark of their own brilliance and individuality. It is quite possible that inventors believed (perhaps not without good cause) that if they once succeeded in hooking a 'typewriter' (as typists were originally called) on their own individual letter order, they would tend to stick to their product rather than re-learn a different keyboard arrangement on a competitor's machine. And they were probably correct.

The component letters of the words A-N-D and T-H-E often tended to be placed next to each other for obvious reasons, but apart from that, logic and reason simply went out of the window. Even consecutive models of the same make did not necessarily offer uniformity (thereby neatly dismissing the theory propounded in the previous paragraph!). On the original World index, for instance, the letter order was:

&CJPFUBLWTHEROIANDGSKYMQVXZ

while on the second model, it had become:

QCBLUIPREANDOSTMYHJFGVXZK?W

and this kind of anomaly was more than common-place.

By the latter part of the century, however, the question of letter order had largely polarized itself into two opposing camps, each with its own preference.

49. Blickensderfer 6 (Courtesy of This Olde Office, Cathedral City, California)

One of these was the 'Ideal' keyboard arrangement which was by far the better and more sensible of the two and which was popularized on such great machines as Blickensderfer (**49, 63 et al.**) and Hammond (**50, 71 et al.**). Common sense featured strongly in the Ideal design, with the ten most frequently used letters of the alphabet placed in the same row—these ten letters account for some 70% of all the words in the English language. The letter order was:

DHIATENSOR

This keyboard order did not just happen—it was designed, and it made good sense…and of course it failed, as any typist today is only too well aware.

50. Hammond (Courtesy of Sotheby's, London)

51. Type Writer keyboard (from the first Type Writer catalogue)

Practice upon the above by touching each letter (one at a time) in any desired word, and the "space-key" after the word. One or two hour's practice, daily, will soon enable you to write from 50 to 100 words per minute, upon the machine. Please call at Office of

REMINGTON SEWING MACHINE COMPANY, 258 West Jefferson Street, Louisville, Ky.

WANTED—EVERYBODY'S COPYING to be done on the TYPE WRITER.

Instead, total acceptance was ultimately achieved by the 'Universal' keyboard, which we all use today when we type. This terribly inefficient design was thrust upon the world by the inventors of the machine which was to become the famous Remington. For reasons which today seem hardly credible, we find ourselves lumbered with this letter order thanks to Messrs Sholes and Glidden (and Remington), who first used it on their machine and, as the accompanying contemporary illustration shows **(51)**, it has remained with us virtually unchanged to this day.

In fact, what we have always been told in the past about the origins of this letter order turns out to be largely nonsense, as we will see in the next chapter.

So much, then, for keys and keyboards as factors which typewriter pioneers had to consider. Bear in mind that first generation machines were being invented almost in isolation—there was no established market for them, there were no experts, there were no typists and a very large proportion of the population abhorred the very idea of dehumanizing correspondence by abandoning script and penmanship. As the second generation progressed, the accumulated knowledge and experience enabled inventors to learn from each other's mistakes while developing methods and systems of their own, different enough from their predecessors (although not necessarily better) to justify their own patent applications.

The next element to consider in detail, after the keys and keyboards, is the means by which the instrument is made to type the desired characters. We have already looked at the use of type bars for this purpose, but the subject embraces a much broader spectrum than that. A number of alternatives was devised and it was by no means certain from the start just which system would prevail.

The best alternative to the type bar, and certainly the most important in terms of the ultimate development of the typewriter, was the type wheel **(49)**. In its broadest sense, which includes just about every conceivable design for locating letters on curved or cylindrical surfaces, the type wheel principle was a serious competitor to the type bar until the early years of this century when it fell by the wayside, only to be resurrected in the IBM 'golfball' with which we are all familiar. The mechanism which operates the elliptical surfaces of this particular design is very sophisticated indeed and although a man called Richardson patented a virtually identical idea over a hundred years ago, it remained a historical rarity until relatively recent times.

Much more common was the use of a regular cylinder for this purpose and inventors over the decades did everything conceivable to this basic geometric shape. They lengthened it and shortened it, they expanded it into a wheel and contracted it into a sleeve, they mounted it vertically and horizontally, inclined it forwards and backwards, they cut larger and smaller sections out of it to use merely an arc, they swung it up and down and they mounted it centrally, to one side, above or below.

Type bars and type wheels were more or less contemporary developments. The first type bar machine was invented by an Italian called Pietro Conti and dates from as early as 1823. He called his invention a Tachigrafo or Tachitipo and it consisted of type bars, hanging in a circle, being thrust upwards onto the paper which was attached to the underside of a flat plate. A keyboard served to select the desired letters, the plate with the paper attached remaining stationary (except for line spacing) while the entire type basket moved a space at a time as typing proceeded.

A remarkable design, well received by all accounts, but premature by many decades. It helped future inventors though, because the design and descriptions were well documented. Within ten years a Frenchman by the name of Xavier Progin virtually turned the whole thing upside down in his Plume Ktypographique. Paper was held in a flat frame, as before, but it was now at the bottom of the machine with the type basket riding over it. There was no keyboard, vertical rods attached to the type bars formed a circle above the instrument and typing was performed by pressing down on these rods causing the corresponding type bar to strike the paper below.

There was one enormous advantage to this concept as is immediately apparent—the paper was beneath the machine rather than on top of it, permitting the typist to read his or her work. This made it the earliest of all typewriter inventions to offer visible copy, and not even Remington offered this feature when they began mass producing machines more then half a century later. The company maintained that you had no more need to see what you were typing than a pianist needed to see the keyboard while he or she was playing, and they dug in their heels and stubbornly stuck to this view until they were eventually forced by the competition to ditch blind machines in favour of the visible-typing upright.

So type bars were firmly established from the start, and whether they struck upwards or downwards, from the front, rear, or sides of the machine, or any combination of these, was really all the inventor needed to design. But inherent in the system were some enormous technical and mechanical problems. Not the least of these was the infuriating tendency for type bars to jam at the printing point, a problem we have already touched upon. It was not so bad if you typed slowly and with one finger, but generally the type bars were no more than pivoted at the extremity (although some later more sophisticated ones were mounted in ball bearings) and tended to become sluggish with wear and dirt, even if they had started out swinging freely. They also had a marked tendency to distort and bend and warp. Neat and correct alignment of the printed letters was therefore a continual problem until some form of guide was developed…and then, the type bars tended to jam in the guide if they were at all out of alignment.

All of which made the type wheel a very attractive alternative indeed. Here was a principle, as its advocates were quick to point out, which suffered from none of these failings. Correct alignment was assured because all the characters were on a common surface and there was nothing to jam up the works—in theory, at least. It sounded like the perfect solution and might have imposed itself from the start except that it suffered from a few problems of its own.

The first of these was the need to provide linear as well as circular motion to the type wheel spindle. In other words the type wheel had to be turned to the desired letter and then had to be swung against the paper. Not an insurmountable problem, it is true, but more than one inventor came to grief over it. Others gave up trying to overcome the problems and took a different approach, mounting the type wheel on a fixed spindle and pivoting the paper carriage in such a way that the carriage was made to strike the wheel. Yet others brought paper and type face together by means of a little hammer, which sounds strange but provided a fine and uniform text in the days when this was no foregone conclusion.

The other obvious problem type wheels faced was that they needed to be turned a different amount for different letters and this seriously affected not only speed but also uniformity of touch. It obviously requires less time and effort to turn a type wheel to an adjacent letter than to one at the very end of the wheel's revolution.

Type bars and type wheels were more or less contemporaries. It was a mere fourteen years after the first type bar invention that a French clock maker by the name of Gustave Bidet produced the first type wheel design. He called it a Compositeur Typographique Universel presumably because he believed it could be adapted for all kinds of printing processes, but its importance lies in its specific contribution to the history of typewriters rather than to the history of printing. Predictably, it incorporated differential spacing which would have been essential for the printing trade. Of course, Bidet was not the first inventor to offer this feature. Progin had already done this, to a limited extent, by allotting more space to capital letters than to lower case, but Bidet was the first to extend the concept to the whole range of typing, even though his differential spacing process had to be performed manually and not automatically by the machine itself selecting the corresponding space for the letter being typed. But Bidet's main importance lies in the fact that he was the first inventor to abandon a flat paper table in favour of a cylinder around which a sheet of paper was fitted.

Thus another element of a modern typewriter—the platen—was duly invented.

It was not of course universally adopted—many inventors still preferred the flat surface—but interestingly enough, many of those who adopted the idea of a platen over the next thirty or forty years used Bidet's concept of typing around the cylinder rather than along its length as on modern machines. After all it is a great deal easier to design an efficient spacing system by rotating a cylinder a given amount after each impression than by moving the cylinder laterally along its axis. There are no real disadvantages to this approach except that it is impossible to read what has been typed without expending a considerable amount of time and trouble but many future inventors on both sides of the Atlantic were prepared to tolerate this inconvenience, or rather, were prepared to force their users to tolerate it. Or perhaps they were simply unable to design a more effective alternative.

By this stage, approximately the 1830s or so, all the various elements of typewriter design are beginning to fall into place. We have the type bars or type wheels which do the printing and the keyboards which control them and select the desired characters. The carriage to hold the sheet of paper is there, as well as the escapement which moves the paper as required. The one essential element of typewriter design still to be considered is how the words are made visible on the paper.

These days we open a new ribbon box and fit the ribbon to our typewriter, rarely considering what we would do if we had to make it ourselves or how we would go about devising a system if none were available. If we stop to think about it, the ribbon would not necessarily be an obvious or automatic choice if we were in the process of inventing a writing machine from scratch, and we might even decide to use it only as a last resort.

Consider the problems. First we have to cut off a long narrow strip of cloth—from a bed sheet, for instance—then prepare what-

ever it is we are going to soak the cloth in, afterwards hanging it up to dry. It must leave a mark even after it is dry, so it will be messy winding this strip of cloth around whatever we have invented, fitting it into the machine, and feeding it through the path it has to take to whatever device onto which it is going to wind. Moreover it has to be kept clear of the paper but at the same time close enough to be useful when needed. Some form of tension or guide system has to be devised as well, and the ribbon needs to be moved after each impression to prevent the type face from drilling a hole in it.

Even if everything goes smoothly and none of these problems occurs, we still have to be very careful, because with all the elements often hidden from sight within the bowels of the machine, we would have to guess or notice precisely when the ribbon reaches the end of its run so that it can be turned around, rewound, or reversed. In fact, reversing the ribbon direction was not the kind of obvious automatic feature one might expect to find on an early typewriter. Even such a prestigious mass produced machine as the Remington did not offer automatic ribbon reverse until almost the end of the 19th century.

How was the problem tackled, then? In almost every conceivable way. It was in the 18th century that the first musical typewriters introduced the concept adopted time and again by later inventors, of marking longer and shorter lines on paper using either sharp points to leave scratches or some form of pencils or pads soaked in ink to trace more visible marks (**25, 26**). Quills were adopted by automata makers, some even with automatic inking, and remained in use in the 19th century in writing and duplicating frames. Extensions of these principles were resurrected in the early years of telegraphy when dots and dashes were automatically drawn on paper tape inside the receiver at the same time as the trans-

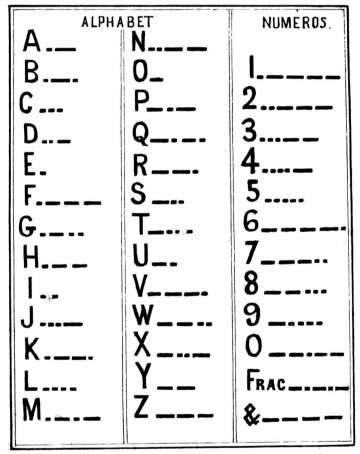

52.   Bain's Electro-Chemical Telegraph code

mitter was sending the message. In fact early printing telegraphs embodied most of the technology devised by typewriter inventors and added some quite unique elements of its own, such as Edward Davy's (and later Alexander Bain's) electro-chemical telegraphs which used electric impulses to decompose chemically-soaked paper at the point of contact, thereby leaving a permanent discoloration in the form of dots and dashes (**52**).

In the case of typewriters, just about everything it is possible to imagine was tried at some time or other. Turri, very early in the 19th century, devised some form of ink applicator which worked badly, as we have mentioned, and also tried 'carta nera' or carbon paper which was a lot more trouble for only marginally better results. As far as we know Turri did not actually invent carbon paper independently; more probably, he had knowledge of Wedgwood's patent which was granted a couple of years before the earliest Turri letter we are able to date. However Turri was obviously working on his machine long before Carolina typed her first letter on it and the earliest letter to have survived is almost certainly not the first to have been typed. So when we date Turri's invention to 1808, it is not strictly accurate—it is simply the date of the earliest surviving letter. On the other hand, we know that Turri was a member of a kind of inventors' club called Società d'Arti Meccaniche and it is possible, indeed probable, that he would have heard of Wedgwood's carbon paper invention through contacts with local and overseas colleagues.

Carbon paper was in use from the earliest days and just what a trailblazer Turri actually was may be gauged by the fact that twenty years later a Frenchman by the name of Gonod, whose 1827 invention was the first stenographic typewriter on record, could do no better than split the carbon paper process into its elements and use them separately. Gonod's suggestion was to insert a greased sheet of paper between the type face and the page so that a greased impression of each letter was typed, later to be made visible by dusting the whole sheet with lamp black and blowing away the surplus. One or two of his alternative suggestions were even wilder, and included rubbing the platen of the machine itself with a mixture of fat and lamp black and placing the sheet of paper over this so that the message would appear on the underside of the page and would be legible by holding the page up against a strong light and reading through the paper.

But more conventional carbon paper was there to stay. In 1836, almost thirty years after Turri, an Englishman called Miles Berry more or less re-invented Creed's musical typewriter and used carbon paper for the impression. Three years later Louis Jérome Perrot in the first of his many patents for a Machine Tachygraphique specified using carbon paper, although it was he who had also suggested the alternative of rubbing the fat and lamp black directly onto the platen. From then on, carbon paper was tried repeatedly: by Wheatstone before he switched to ribbon in later years, by Beach before he switched to embossing, by Foucauld who used it in the 1850s, and by Pratt who was still specifying it as late as the 1860s—the list goes on and on.

Despite the fact that Alexander Bain actually invented the modern typewriter ribbon as long ago as 1841, it was not until well into this century that it became standard typewriter equipment. Inking pads and rollers were strong competitors, and their intrinsic features made them obvious choices for those who designed alternatives to the type bar principle. But other alternatives abounded, including the use of pens, which, like carbon paper, were by no means confined to the 18th century—a Frenchman, Antoine Dujardin, was still advocating their use as late as 1847.

Dujardin had first appeared on the scene in 1838 with an instrument he called Tachygraphe which was itself far from original. It had a piano keyboard with the twenty-six keys in alphabetical order and its twenty-six pens printed a series of dots on paper tape advanced by clockwork…*déjà vu* indeed, for Creed and Unger had designed a virtually identical machine almost a century earlier.

But Dujardin's purpose was not to type music. Each letter produced a dot in a specific location across the width of the paper tape. By running the tape beneath a scale with the letters of the alphabet on it, the location of the dots would indicate the letters which had been typed. This principle was to be used on more than one occasion in the years to come, and most specifically in cryptography, for by scrambling the letters on the keyboard consistent with those on the scale, a primitive cryptographic method would result. But clearly, Dujardin was not likely to have much success

with a machine like that, and a modified version using columns and coloured inks instead of the scale was also doomed to stay on the drawing board. He kept trying however, remaining loyal to his pen and ink principle in the printing telegraph which he went on to patent in 1847.

Meanwhile, inking pads and rollers were firmly established as serious competitors to carbon paper and ribbons—the very characteristics of a turning type wheel made it ideal for matching with an inking roller. The system required that the roller be kept moist and turned so that a consistent supply of ink was always on offer. Bidet, whose Compositeur Typographique Universel has already been discussed, was the first to marry these two elements together in 1837. The American Charles Thurber specified inking rollers a few years later, as did Fairbank, Cooper, and many others until the years of mass production when such ubiquitous machines as the Blickensderfer **(49, et al.)** confirmed the bond between type wheel and inking roller beyond any doubt.

The exceptions to this generalization are numerous. Not only were there some interesting combinations of type bars with inking rollers (the front stroke Sun **(8)**, for example) but there were also many combinations of type wheels and ribbons (the Frolio **[6 (r)]**) and type sector and ribbon (the Hammond **[50]**). On the other hand, inking pads, which were ideal for index machines, were also holding their own in special type bar applications from the pioneering Progin's invention to the sophisticated later machines of the so-called 'grasshopper' design such as the Yost **(53, 60 et al.)** and Williams **(54, 76 et al.)**. But those are the exceptions. By and large, type bar designs used ribbons, while Bidet's 1837 combination of type wheel and inking roller established a historical trend which most of this ilk were later to follow.

A Detroit resident called William Austin Burt had already started a different trend of his own some eight years earlier. Burt patented his Typographer in 1829 in the very infancy of the typewriter as a potentially practical appliance and although the original instrument has not survived, correspondence typed on it is eloquent testimony to the quality of work it was capable of performing. As can be seen from the accompanying illustration **(55)** of a reconstruction of Burt's invention, the Typographer was of swinging sector design, with the type face in an arc fitted to the underside of a bar pivoted at the far end of the machine and finishing in a point at the front where it swung over a curved index, selecting the desired letter and thereby bringing the corresponding type over the printing point. Pads on either side provided the ink. The paper was held flat in the machine and a clock face on the front geared to the paper frame helped the typist keep track of lines and spaces. The machine typed upper as well as lower case—the first ever machine to do so, about half a century before its time.

This was totally different to the previously mentioned Bidet's effort, and on the face of it not nearly as good, despite the fact that it typed both upper and lower case. Burt's was the earlier invention by a few years, but not enough happened in the years between Burt and Bidet to attribute the difference to learning from others' inventions. It was Bidet's conception that was different and better, as was its application to the problem of designing a typewriter. However, both designs spawned a myriad progeny during the rest of the 19th and the early part of the 20th centuries, many of which died in their infancy while others developed into fine and full maturity.

Before we move on from these early developments we ought briefly to consider two further features which were devised in the earliest days and have remained with us ever since. Both relate to the means by which the text is applied to the paper. We have already examined many of the possible alternatives, which ranged all the way from simply scratching a mark on the paper to leaving an electrolytic image. Of course the idea of scratching the paper

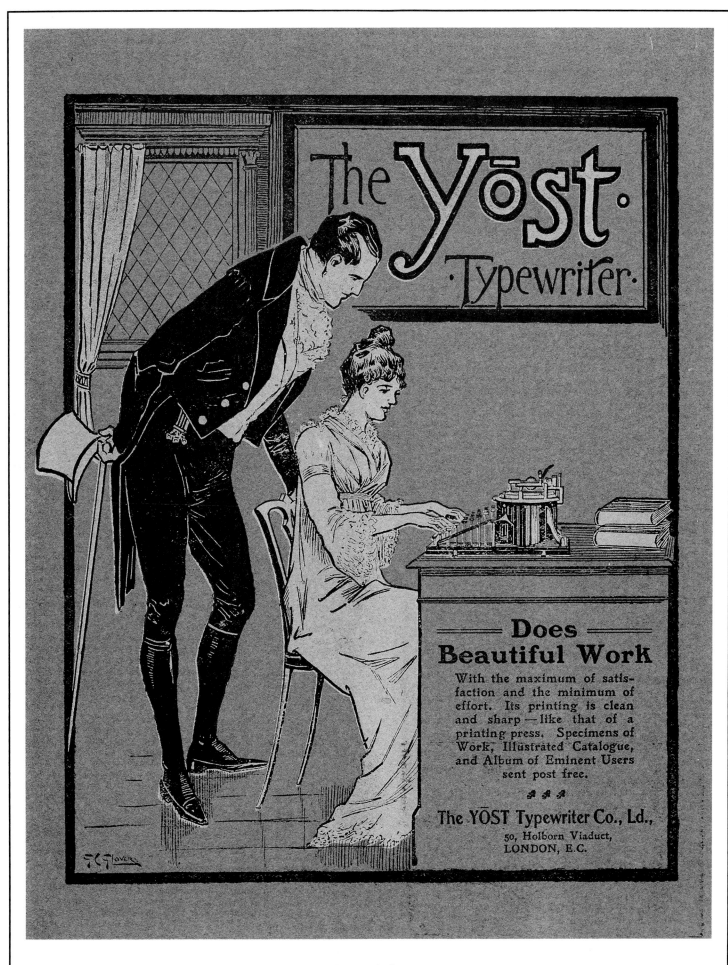

53. Yost leaflet

54. Williams

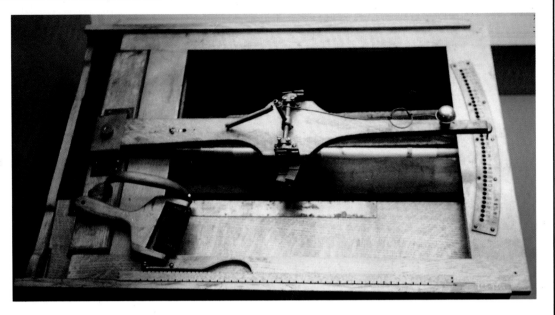

55. Burt (Courtesy of Dan Post Archives)

was primitive, but it was only a relatively small step from scratching to embossing and this opened up a whole new field for the writing machine as an aid to the blind.

Carolina Fantoni, it will be remembered, was our first blind typist in 1808 or earlier. Just how she managed is hard to imagine, for as far as we know from her correspondence **(35, 36)**, her machine produced no copy which she herself could read. Maybe she was endowed with an amazing memory, for she typed slowly and laboriously with arthritic hands and was plagued by headaches, yet made virtually no mistakes in a page of typing which must have taken her a long time to complete. Or perhaps she had help from a sighted relative, friend, or servant who could read back for her what she had written if she were interrupted or lost track of her thoughts. It seems remarkable that she was able to correspond at all, and yet correspond she did; compared with other early typists, is it just that the others were less proficient, or are we simply ignorant of how Carolina Fantoni did it? We may never know the answer.

We do know that other inventors of machines for the blind faced a similar and very real dilemma on this score. Essentially, they had to decide whether their inventions should leave a tangible image capable of being read by the blind typist or recipient, or whether they should merely type a visible message capable of being read by the sighted. Most opted for the former alternative. A few offered both.

Embossing the message was nothing new by the 19th century. Henry Mill was in all probability alluding to it in his 1714 patent, when he specified that his machine was capable of *'impressing'* its copy *'deeper'* than any other writing. Mill was concerned with forgery rather than blindness, though, so the obvious extension of the embossing principle for use by the blind remained temporarily undeveloped.

Over a century later, the idea surfaced again in an even more improbable guise, and once again its importance to the blind was overlooked. It was in 1835 that a certain famous American gent by the name of Samuel Morse, who is of course popularly credited with the invention of the telegraph, was first reported to have revealed his historic device which embossed its message of dots and dashes on paper tape within the receiver as it was being fed through the transmitter.

Yes, the Morse machine originally embossed…but then he had the even more brilliant idea of dropping any thought of printing or embossing in favour of the famous buzzer with which he is universally associated.

So these were the inauspicious beginnings of embossing as an alternative to printing, and it was only some twenty or so years later that the next decisive step was taken, but a significant step it was. In 1856 one of the editors of the *Scientific American*, Alfred Ely Beach, patented a typewriter for the blind which embossed on paper tape. Whether this feature was the result of circumstance or design remains speculative, for nine years earlier Beach is known to have made his first abortive attempt at inventing a typewriter which used carbon paper for printing. This earlier effort was not a success, however, and it is possible that the carbon paper may have been at least partially responsible for its failure, if the design of the later machine is any indication. Of course, we have seen time and again that carbon paper was acceptable enough to many an early inventor and achieved significant results, but Beach was not satisfied with it—either the machine was at fault or he was more critical than others.

His original 1847 design, which featured a respectable keyboard and basket of type bars, might well have proved successful in the end had Beach but persevered with it, but his second effort was probably too complex to stand a chance. Instead of simply perfecting the single circular basket of type bars, which might have allowed him to type on whole sheets of paper, Beach used a double type basket—one on top of the other—so that when a key was pressed down the two corresponding type bars were activated simultaneously. One came down from above and struck its mate coming up from below. One of the two type faces was male and the other female, so that what was produced was a raised letter where the paper had been clamped between them.

It was a very complex solution to the problem but need not be condemned on that score alone except that, by opting for such a configuration, Beach was forced to abandon any hope of typing on a sheet of paper because there was simply no place to put it. He had to settle instead for a thin strip of paper tape wending its way between gaps in the type bars. That may well have doomed the design even more than its mechanical complexities.

However, as we have seen repeatedly, many otherwise abortive efforts made significant contributions to the developmental history of the typewriter in the end, and Beach's was no exception. Embossing the text was important for the blind, of course, but here he was merely adding his voice to the chorus of those who went before him. What is more important is that his invention looked precociously modern, with its three row keyboard of small cylindrical keys worthy of any early 20th century design, let alone 19th century prototype. Furthermore, as well as this impressive exterior, there lurked within the bowels of the machine an ingenious escapement using a form of universal bar, which was destined to become adopted as standard equipment by generations of later typewriter manufacturers.

Even the process of clamping the paper between male and female moulds of each individual letter was destined to live on after Beach's second invention in 1856. Five years later, the Brazilian priest mentioned at the beginning of this book—Padre Francisco João de Azevedo, to give him his full name—burst onto the scene with his stenographic machine (23) which received quite a rapturous local reception. Originally conceived as a typewriter for standard printing, it was only developed as a shorthand machine to satisfy the requirements of the authorities who subsidized its development. Maybe they understood more than the inventor himself, for it is difficult to imagine how the design would have succeeded as a full typewriter. It printed its shorthand message on paper tape with larger or smaller gaps between the component letters of a word, which of course is quite acceptable with stenography, but not with text. The system was a combination of a fixed plate and type bars on which there were male and female characters. The paper tape moved across this plate and when a key was pressed the corresponding type bar clamped the paper between bar and plate, creating the embossed character. There was therefore no common printing point—the letters were simply embossed in their correct sequence but with whatever gaps resulted. It was this unfortunate feature which would have proved a serious handicap to the use of the machine for regular typing.

The embossing process was never a serious contender in typewriter history, other than for the blind, but it was not as abortive as might at first appear. Those of us who have settled uncomfortably into old age will have little trouble recalling those coin operated machines which adorned railway stations when we were kids (19) and which, for a penny, would permit us to print our names (or other equally important secret messages, such as *Michael loves Helen*) onto a short strip of aluminum, conveniently curved and cut away at the ends no doubt to allow us kids to tack them onto strategically located trees or telegraph poles. These nostalgic and evocative devices were writing machines without any doubt, their origins easily traceable to the many 19th century patents covering similar circular index designs. My own obsessive childhood fascination with these machines, whenever I could lay my hands on a penny, was undoubtedly a significant contribution to my later interest in historical typewriters.

Of course, embossing has appeared again even more recently. A new generation of writing machines has developed with the small hand held devices, like the Dymo, using a circular index and embossing self adhesive plastic tape. Some of these modern instruments are the design, size, and shape of the 1887 Pocket typewriter or the Taurus (56) but without the inking roller, and embossing instead of printing. Many such comparable gadgets are currently available, their roots extending deeply into typewriter history.

We will return to embossing shortly when we examine printing telegraphs. In the meantime, there is one other method of leaving a message on paper which we have not yet discussed and that is the not unrelated process of perforation. It is a far from desirable alternative to the printed impression where the image is thoroughly visible and it must be said that the few inventors who opted for perforation did so *in extremis*, when all else failed, or else they began with perforation and dropped it as soon as they were able to devise a better substitute. However, if you share my personal view that nothing in the history of invention is ever totally wasted, then you will be gratified to know that even this idea was not without its merit and subsequent application.

Foucauld (16), as we have seen, appears to have been the first inventor to specify perforation back in 1839 and he was still using it in his own later development of his machine. However, only one other man really applied the idea and that was the Austrian Peter Mitterhofer (24), a quarter of a century later. Mitterhofer belongs to those typewriter inventors who invented virtually nothing new themselves, but who combined the design features conceived and developed by others into more or less practical prototypes of their own.

Less rather than more practical, in the case of Mitterhofer's four models. The first dated from 1864, but it was only on his last model, five years later, that he specified an inking system and this he based on a unique combination of feathers and bristles. So anachronistic was the design that it barely warrants a mention particularly as the inventor, who had been a poor man without resources when he had built his first three instruments, had meanwhile received some financial patronage and should have been capable of better. On his earliest efforts, he had simply embedded tips of needles into the ends of the wooden type bars and perforated the page with them.

Yet at roughly the same time as Mitterhofer was whittling away at the pieces of his device, other inventors were already grappling with advanced notions of automatic telegraphy. Despite suggestions from some quarters to the contrary, it could hardly be claimed that they were in any way indebted to Mitterhofer. He chose to make holes in paper because it was the best—the only—process he could devise at the time, even though it was totally inadequate for the purposes he had in mind. In the case of the printing telegraph, however, the use of perforated paper was a deliberate and conscious development and was a brilliant solution for the problems to which it was applied.

Essentially, the trouble with telegraphy in its early days stemmed directly from its great impact upon human existence. Such were its massive applications and repercussions that it had already outgrown its capabilities almost immediately after its invention. There was no gentle, gradual development and growth over a century or more for the telegraph as there had been for the typewriter. The telegraph quite simply revolutionized civilization from the very beginning.

Communicating with someone who was out of sight and hearing was itself not a new obsession—it had possessed mankind from the very dawn of history. It is not difficult to imagine that the man who first tamed a horse and rode it to his destination must have felt that he had achieved a giant step forward in communicating with his fellow man. There are parts of the world—such as the Andes in South America—where horses could not be used and where human runners carrying coded messages were the only means of communication until very recent times indeed. Elsewhere, the terrain being flatter and more suited to visual methods, messages could be transmitted over relatively short distances by means of lights or mirrors, in preference to horseback. Smoke from fires was successfully adapted by the North American Indians while other cultures harnessed the homing instincts of certain animals such as pigeons to achieve communication over even greater distances.

This was the state of the art for millennia until the potentials of electricity were first examined and applied to the problem in the late 18th century. Claude Chappe (who is credited with coining the word 'telegraph') devised a form of mechanical semaphore in 1792 using a post with movable arms, but such visual systems were quickly displaced as the characteristics of electric magnetism were found to be so ideally suited to the purpose. Close on the heels of Alessandro Volta's historic development of the storage battery, the properties of this new-found source of electricity were being applied to the decomposition of liquids, and from there to the permanent discoloration of paper soaked in these liquids.

That these properties of the electric fluid could be applied to telegraphic printing was not lost on early inventors, and numerous systems were soon suggested. Long after the 1830s, when Samuel Morse immortalized himself with his telegraphic embossing invention using a combination of dots and dashes for the letters of the alphabet, the electro-chemical telegraph seemed set to dominate the field. Commercial applications had already been initiated on both sides of the Atlantic. There was every reason to suppose that the method would succeed, if not prevail.

Edward Davy was the first to develop an electro-chemical telegraph in 1838 by decomposing chemical compounds on a piece of cloth by means of an electric impulse which left a permanent mark on the cloth. But it was Alexander Bain, who had already chalked up a few impressive patents for printing telegraphs, who came up with a truly remarkable device which he patented in 1843. Bain was an impecunious clock maker of considerable skill and he already held a number of patents for electric clocks and telegraphs which the great Sir Charles Wheatstone is arguably believed to have plagiarized.

Bain's 1843 electro-chemical invention, however, was his and his alone. This ingenious device consisted essentially of a weight driven clock with a feeler attached to the pendulum in such a way that as the pendulum swung, the feeler scanned the surface of the weight. With each swing of the pendulum, the escapement of the clock would release the train of wheels by a small amount, thereby dropping the weight and presenting a new line for the feeler to scan. The weight itself was insulated but the inner surface was live and so was the feeler; whatever contact the feeler made with the surface of the weight was transmitted as an electrical impulse onto an identical synchronized machine serving as the receiver, in which the surface of the weight was covered by chemically treated paper (sodium nitrate and potassium ferro-cyanide were used) which discolored whenever the feeler was electrically charged. In this way, text, code (52), and even pictures could be readily transmitted.

In Britain, Bain's device was used for ten years or so but in the United States injunctions from rival Morse interests claiming patent infringements prevented it from getting off the ground.

Morse's scheme was better than Bain's anyway, and no doubt would ultimately have prevailed even without the injunctions. What Morse had done, back in 1835 when he later claimed he had first shown his invention to some friends, was to send a programmed message from a transmitter to a receiver resulting in an embossed line of dots and dashes on paper tape advanced by clockwork. He originally devised his arrangement of dots and dashes as points and flats on the edge of a slide; when this slide was passed through the transmitter, a follower opened and closed the circuit as it rode over the contours of the slide, causing a stylus in the receiver to emboss the paper tape correspondingly. Initially it transmitted a code of numbers which referred to words in a telegraphic dictionary. However, within a remarkably short period of time, Morse abandoned embossing, made an attempt at printing the message and then dropped that idea, and ultimately settled on the code and audible signal with which his name is linked.

I said Morse 'claimed' to have shown his invention to friends in 1835. He also claimed he first thought of it in 1832, although his patent application was not filed until 1837 and the specifications in 1838. The patent was not actually granted until 1840. Those eight years between 1832 and 1840 may seem a very short period when seen in a cosmic perspective, but in those hectic times, days and weeks, let alone months or years, were often the difference between immortality and relative obscurity, between wealth and poverty. Without diminishing the stature of Samuel Morse in any way, it must nevertheless be noted that priority of invention is attributed to him on the strength of retrospective claims of questionable impartiality, and at least two other contenders for the honours might

well have had good reason to feel aggrieved. One of these was an Englishman by the name of Alfred Vail who in 1837 invented a printing telegraph using a vertical wheel with twenty-six type plungers to type the letters of the alphabet, rather than code, onto paper tape—an important if little heralded innovation. The following year (while Edward Davy was putting the finishing touches to his electro-chemical device) a German, Professor C. A. Steinheil, invented a printing telegraph using twin inking styli to produce his own code of dots geometrically juxtaposed in two rows, with clockwork to advance the paper.

So once again, in the case of the printing telegraph as well as the typewriter, the essential components were more or less in place well within our second generation. Only one major element was still needed, however, and that was the one needed to bridge the gap between transmitting and printing codes and doing the same with plaintext.

Basically, what was required was a typewriter which would type out its message at the same time as it was transmitting it to a receiver simultaneously typing the same message—something with a keyboard, a printing device, some form of paper carrier and an inking system. Vail put together his first designs for just such an instrument in 1837 using type plungers fitted to a wheel in such a way that they struck radially outwards against the paper. Neither this machine nor the subsequent modified versions were a success, but it was the first such invention. However, it was left to two other inventors—contemporaries, rivals, and enemies—to fight for the greater honours in 1841.

The inventors involved were Charles Wheatstone and Alexander Bain, whom we have already mentioned in connection with the electro-chemical telegraph. Today, Wheatstone is undeniably the more famous of the pair and overshadows the unfortunate Bain, whose contributions had remained virtually unheralded until revealed at some length in *The Writing Machine*.[1] Until then, he had been ignored by virtually every original source except Martin, who nevertheless dismissed Bain in a mere two lines.[63]

There are several reasons for this, not the least of which is that while Bain's actual machines are lost, Wheatstone's prototypes have survived in the London Science Museum collection where they occupy their rightful pride of place. The other reasons may have something to do with Wheatstone's personality, which seems to have induced him repeatedly to assume credit for the efforts of other lesser mortals with whom he was associated. It is a significant but oft forgotten fact that even the most famous of his contributions to modern technology—the so-called Wheatstone Bridge—was actually the resistance bridge previously invented by a certain S. H. Christie.

It appears that Bain, who was hard-up and unknown, was introduced to the famous Wheatstone and proceeded to reveal to him his own thoughts and plans relating, amongst other things, to printing telegraphy. Bain also showed him details of his electric clock invention…and sure enough, Wheatstone secured a patent on the device a few months before Bain, although he subsequently lost the legal battle over it.

Wheatstone meanwhile was presumably involved in a separate legal wrangle with another of his associates, William Fothergill Cooke, over some patents relating to telegraphy. Bain had nothing to do with this particular affair, although an advocate of Bain's who wrote a book about him believed the dispute was a trumped-up affair inspired by Wheatstone to gain publicity for the invention. The book was by John Finlaison, Esq. who was an 'Actuary of National Debt Office, and Government Calculator' and therefore a worthy gentleman and, one assumes, a man of some substance and credence. The unabridged title of his book states its case in no uncertain terms: 'An Account of the Remarkable Applications of The Electric Fluid to the Useful Arts, by Alexander Bain;

with a Vindication of his Claim to be The First Inventor of the Electro-Magnetic Printing Telegraph and also of the Electro-magnetic Clock.'[33]

That briefly was the scenario. Bain's 1841 patent protected two very different designs for printing telegraphs. One was a device which used a separate electrical circuit and electro-magnet for each character with a circle of eight centrally located keys controlling type bars for printing on paper tape. This was far too complex for practical purposes. The other device was based on a type wheel and was spring driven, using a single line with earth return for the electrically controlled escapement only. Two years after this, his remarkable chemical telegraph recorder made its appearance.

There was more to come from Bain, but it was of lesser importance even though he was able to enjoy some moderate success with his inventions later in life. However, he will probably best be remembered for inventing the typewriter ribbon in his 1841 patent by suggesting that a mixture of oil, lamp black, and turps be rubbed onto a roll of ribbon ½ inch wide. Two years later he realised that merely soaking the ribbon in printer's ink and drying it was sufficient, and he duly patented that improvement. Finally in his 1845 patent, he made it a silk ribbon and wound it between two spools.

So there is yet another element of the typewriter in place. Like everything else, it was not to be universally adopted until well into this century, when typewriters were being mass produced in millions and ribbons could be bought at virtually any corner shop. In the beginning, however, it was an incredibly messy process, making your own ribbon, soaking it, and hanging it up to dry. Unforgettable descriptions of the agony of it have been handed down to us by the early pioneers and with good reason, as anyone who tries to follow their instructions will discover.[93]

Wheatstone meanwhile was at work developing the original 1841 patent along lines very similar to Bain's at first, before branching out into a different design inspired by yet another associate. The principle difference between the two patents was that instead of using type bars or wheels, Wheatstone specified a form of what we have come to call a daisy wheel, with carbon paper and a cylindrical platen. Ten years later, whether on his own or (more probably) once again with associates, he had greatly improved the design, fitting a piano keyboard and locating the type on the ends of the flexible teeth of a comb of swinging sector inspiration. Inking was by pad, typing directly onto tape thereby dispensing with the need for a platen, but only temporarily.

On Wheatstone's next effort, dated 1856, jointly with a Mr. Pickler, the platen was back, small cylindrical keys replaced the piano keyboard and the whole thing was turned on its end so that its profile was vertical rather than horizontal. The type on a comb and pad inking were retained. A final fling some years later was not blessed with any particular advantages; one has the impression that the great man had lost his way a little (or perhaps run out of associates to plagiarize) and deviated instead of concentrating his energies on perfecting one or other of the more promising of his earlier designs. By then, Wheatstone had moved on from his original interest in the printing telegraph and that was a serious mistake on his part, for while the typewriter struggled in relative obscurity in the course of those decades, telegraphy was shooting ahead by leaps and bounds.

56. Taurus (Courtesy of the late Josef Zimmerman)

One final short note of interest about Wheatstone before we proceed. In 1854 he claimed the honour of inventing a simple enciphering device which he called a Cryptograph **(57)** and which was first displayed at the Paris Exposition Universelle in 1867. True to form, this appears not to have been an original Wheatstone invention, either. The device consisted of two concentric circular indices, the outer consisting of twenty-seven segments and the inner of twenty-six. Printed on the outer circle were the letters of the alphabet, in alphabetical order, plus one blank. Written by hand on the inner segments were simply the letters of the alphabet, scrambled at will by prior agreement between the interested parties. One larger indicator pointing to the outer circle was fitted to a wheel of twenty-seven teeth which, by means of a common twelve-tooth pinion, was geared to a wheel of twenty-six teeth to which was fitted a shorter hand indicating the letters on the inner circle. The mode of operation is perfectly clear and simple: each revolution of the larger hand selecting plaintext letters from the outer circle progressively enciphers the letter selected and indicates the encoded letter on the inner circle. The party receiving the enciphered message performs the identical operation in reverse in order to decipher it. Clearly, the system was a little better than that of a simple substitution cipher by virtue of its progressive encipherment, and applications were to find their way into early printing cipher machines for some time to come, but to be honest, as an enciphering system it was all but worthless since such primitive elementary codes were vulnerable to being cracked even by dedicated schoolboys. One has only to consider the encipherment potential of the Hagelin Cryptos **(167)**, not to mention the fabled German Enigma machine **(158)** to realise how primitive and inadequate was the system employed in the Wheatstone Cryptograph.

Telegraphy, meanwhile, had not only caught the imagination of the public but was soon outstripping the potentials of the system itself. Speed was of the essence, and even the very best of human endeavours proved hopelessly inadequate. Messages were technically capable of being transmitted far more quickly than anyone could operate a morse key manually, of course, and they needed to be transmitted at ever-increasing speeds if the demand was to be satisfied.

It was reliably estimated at the time that the maximum potential of the printing telegraph was between 450 and 650 words a minute. To achieve this would have meant the morse relay operating in the region of 160 times per second which was deemed to be the very limit of its capability. Of course, no human operator could ever hope to attain anything even remotely approaching this figure by manual transmission.

The answer ultimately lay in the automatic telegraph. We are looking ahead a little at this point but, unlike the typewriter, the telegraph made itself well nigh indispensable to society from the start and progress was impressive. It did not take long before telegraphists were perforating text onto paper tape at their normal typing speeds and then feeding the accumulated tape through the transmitter at the maximum speed of which it was capable. One machine and one telegraph line could thus actively employ a battery of operators, all manually punching holes into tape on their own perforators so that before long, the original manual transmission speed of twenty-five words per minute was almost contemptible.

Speed was not all, of course, for once the information was punched onto tape, it was there more or less forever. It could be passed again and again through different machines on different lines, as indeed it was in early transmissions of news stories, and it could be stored and retrieved at will.

Thus the seeds of modern electronic and computer technology were already being sown. At one time, it must all have seemed as miraculous as voice writers and voice recognition systems seem today—more perhaps, for in our lives we have learnt to confront technological miracles these days with a nonchalance bordering on boredom. The idea of dictating directly into a typewriter which simultaneously types the text is now no longer in the realms of fantasy and science fiction these days, as it was back in the last century when a Mr. A. C. Rumble of San Francisco or a Dr. Frank Traver of Racine, Wisconsin, tried to achieve just that with their 'phonographic' typewriters.

What would Creed or Unger have made of it all? Or Turri, or Progin…?

57. Wheatstone's Cryptograph

Chapter Three—
Third Generation: The
Age Of Mass Production,
Part 1

'I think I have the right idea at last.'[72] So said Christopher Latham Sholes of Milwaukee to his friend and neighbour Henry W. Roby some time during the year 1866. He certainly was not the first man in history to have cried *Eureka!*, and he was probably not even the first typewriter inventor to have done so. In fact, Sholes was not really the first in anything at all, except that in his case the statement was recorded, as well as the circumstances in which it was uttered, and so it was destined to assume prophetic significance then and now.

It seems an appropriate way, therefore, in which to open the account of the third and decisive generation of typewriter invention which heralded the beginnings of the machine's mass production. In the previous chapter we examined in considerable detail the earlier development of the invention. Those few decades saw every major component of the writing machine invented in one form or other, often many times over. We also examined the process by which these inventions matured into the first tentative attempts at manufacture.

What was there still left to do, then?

Plenty, as it happens, because getting the typewriter out of the patent office and into the limited manufacture of shorthand machines or printing telegraphs or machines for the blind was one thing, but breaking into mass production of a conventional typewriter and ultimately getting the machine into every office and home was to prove a different proposition altogether.

There was no direct logical sequence to events, whichever way one examines them. Being first did not guarantee priority or success, nor did being the best. Rather it was the concurrence of all kinds of propitious omens which did the trick. Boundless energy and a thick skin helped, together with unlimited servings of luck. If you had all these, plus a bright idea or two, plus access to capital, you were possibly on the right track. And if you happened to be in the right country, at the right time…well, Morse, Bell, Edison, and Marconi (to name but a few) all proved it could be done.

But success was evasive if you were short on any of these vital ingredients. Giuseppe Devincenzi might well have reflected on this when he was granted the world's first electric typewriter patent way back in 1854. Today, the arrival of the electric typewriter into our lives is such a recent event that it hardly needs mentioning because we all remember it clearly enough. The intriguing question is why it should have taken almost a hundred years, from the first invention of this kind of instrument, for it to achieve success.

Why indeed! For Devincenzi's was a great idea which deserved a better fate than the ignominy of neglect. In his patent, he applied what was by then the familiar technology of the printing telegraph's comb and pinned cylinder to an ordinary typewriter. It was a good concept and one which, as we know, was to achieve much fame and success in the early stages of telegraphy when comparable technology was applied to synchronized systems in which instruments at opposite ends of the line had type wheels turning in unison. Devincenzi's ingenuity deserted him somewhat when it came to designing an inking system, however, for the best he could de-

vise was a sort of endless belt running through a trough of ink with a wiper keeping the device more or less under control.

Despite this, the world's first electric typewriter patent was a major breakthrough and deserved a better fate, especially when one considers that it was to be more than half a century before the first electric typewriter was (unsuccessfully) manufactured. And it was another twenty or more years after that before the first electrics were successfully marketed, and even those were manuals clumsily fitted with an electric motor which used power only for the carriage return.

Being first, then, was as little comfort to Devincenzi as it was to so many of his predecessors and indeed if this were not an almost universal truth, then it might not have been necessary to re-invent the 'golf ball' typewriter all over again in our own day, a century or more after it was first devised.

We have already seen that the second generation resulted in actual manufacture, and although there are still some important names to consider, by and large the inventing of the machine was a second generation pre-occupation. By the time we reach the third generation, then, we find that even manufacture is of little significance because mass production is the dominant issue. There is a distinction to be drawn and an important one at that. For Foucauld (16), Hughes (2, 44), and Jones all manufactured but can hardly be said to have mass produced. That was to come some twenty-odd years later, as we shall see. But of course it is impossible to point the finger at a specific year in which one phase ended and the next began, because there was bound to be an overlap. Some border-line cases revealed inventors patenting machines and building prototypes which did little more than draw upon the ideas of others at a time when mass manufacture was already a reality. Others were still patenting obsolete and anachronistic designs decades out of period. A few achieved merely limited manufacture well into the age of mass production, although there are some important personalities among them.

First of all, there are a few inventors in this period who never actually succeeded in manufacturing but who warrant separate consideration. Some of these have already been mentioned briefly in other contexts and their names will be immediately familiar.

Charles Thurber is perhaps the most important man we have yet to consider. He was granted his first patent in 1843 for a design based on type plungers arranged vertically around the periphery of a horizontal wheel above a flat paper table (40, 41). This was not the first type plunger idea by any means, but it was an excellent design, and with the added feature of differential spacing, it represented a sophisticated invention for its relatively early date. Even better was the fact that the differential spacing was performed automatically as the typing progressed—in previous inventions such as that of Bidet, the letters were allotted their corresponding spaces manually, in a separate operation. This was not Bidet's only idea which Thurber was to use, for in his second machine he abandoned the flat paper table in favour of a cylindrical platen of very precocious design indeed, even though it typed around its circumference rather than along its length as on modern machines. This is a feature we find time and time again on early inventions, as we have already seen, for reasons which have already been examined.

There was still more to come from Thurber. In 1845 he came up with a fanciful design for a machine which actually reproduced handwriting. Now this Mechanical Chirographer, as he called the device, was an interesting development for several reasons. Often dismissed as retrogressive since it harks back to the 18th century writing automata rather than forward towards new horizons, it is nevertheless a fascinating development because it provides us, by implication, with a graphic verification of the problems early typewriter inventors invariably faced. Apart from the few fanatics who were actually designing and building them, no one even wanted a typewriter to begin with, and this resistance to the machine persisted until well into the last quarter of the nineteenth century. The uses of writing machines in telegraphy and for the blind were recognised, but by and large resistance to typewritten correspondence predominated. People were often offended by it and even refused to receive or acknowledge it. The conventional wisdom was that letters were intended to be written by hand—that personal touch was considered vital. Beautiful script was what counted, not some new-fangled machine-made text.

So Thurber, having invented two progressive typewriters with many worthy and valuable features, unexpectedly changed tack completely and created the machine which produced handwriting. Having used a cylindrical platen to such good effect in his second device, he suddenly abandoned it in favour of a flat vertical surface holding the sheet of paper. You are almost tempted to feel that if you enclosed the whole thing in a globe with just the pencil sticking out you would be back to von Knauss all over again. A hundred wasted years…maybe.

That the Mechanical Chirographer was a flop is beside the point; that it was intended as an aid to the blind is equally irrelevant (and it is difficult to understand why this particular machine should have been better for the blind than either of his first two inventions). The interesting and revealing point is that he abandoned typing in favour of a retrogressive monstrosity which was doomed from the start. Given the climate of the time, he gave up mechanical typewriters, which did not work and no one wanted, in favour of a machine which reproduced the handwriting which they apparently did want.

Another man who worked as hard as anyone on the typewriter was the important Italian inventor Giuseppe Ravizza. If sheer dedication and determination were to be justly rewarded in this world then Ravizza might well have found himself among those most deserving the honours. In the event he was to die a broken and disillusioned man, late enough in the 19th century to witness the beginnings of the Remington success where his own efforts had ended only in failure.

Ravizza's inventions spanned a period of some forty years from his first model in 1847 to his sixteenth machine just prior to his death in 1885. In fact his earliest involvement with the typewriter dates back even further, to 1832 or so, when he began corresponding with his compatriot Pietro Conti whose Tachigrafo is sometimes credited with having initiated the Ravizza obsession. Conti had by then abandoned his 1823 invention and may even have handed it over to Ravizza lock, stock, and barrel, for the first 'Cembalo Scrivano,' as Ravizza called his own machine, was almost a copy of Conti's. Certainly both were type bar instruments of up stroke design—Conti can claim credit as the inventor of that principle, which was to receive such prominence in later decades—with a small keyboard at the front and a flat paper table above.

Three more prototypes were to follow before the design was formalized in a patent granted in 1855 (15). The invention offered few, if any, startling innovations. It had a piano keyboard, up stroke type bars, flat paper table, keys arranged in alphabetical order, silk ribbon rubbed with fat and graphite plus a spring wound escapement. The one major departure from Conti's design was that the type basket remained stationary and the carriage was moveable—this had already become fairly standard procedure in the intervening years.

Later models followed this design with few modifications until his 13th in 1879 when Ravizza adopted current trends—already introduced by Remington the previous year—leaving the innards of the machine exposed rather than enclosed, which made for more silent operation since fully enclosed machines tended to reverberate and amplify the noise. At the same time he realised, well ahead of his American rivals, how important visible typing was destined to become and this 13th model offered the feature of visibility almost thirty years before Remington eventually bowed to the inevitable demand for it.

Ravizza's final effort—his model 16 of 1882—might well have proved a turning point in typewriter history, for at last he had a promoter willing to manufacture the machine in competition with Remington, but unfortunately the plans were aborted in the early stages and all hope of manufacture was abandoned.

Few inventors in typewriter history have left a legacy as rich in dedication as Ravizza, but there are two others in this period who were comparably, if not equally, doomed. Some inventors were at least able to enjoy a modicum of success to compensate them for their efforts, but the unfortunate Samuel Francis of New York and John Pratt of Alabama proved as unfortunate as Ravizza.

Dr. Francis has caused quite a controversy over the years, for his 1857 invention appears to have borrowed heavily from Ravizza's, rather as Ravizza's borrowed from Conti's. But as we have already noted, the second generation was characterized by few totally original inventions: the first generation took good care of those. So yet another up stroke type bar machine enclosed in a wooden case with a piano keyboard at this relatively late date was not likely to cause much of a stir, except that the Francis machine (39) had a few other ideas worthy of mention. One of these was a novel arrangement by which the type bars were flicked up against the platen, there being no direct permanent linkage between key and type bar.

This may not sound like a particularly good idea until one examines the problems caused by the direct linkage on other inventions both earlier and later, when one realises that Francis had devised a very sensible and practical alternative indeed. This was further improved by a device which prevented jamming if two keys were accidentally pressed at the same time, as may happen quite easily with a piano keyboard. However, Francis lost his way a little when it came to specifying inking methods, suggesting a number of relatively clumsy alternatives. What redeemed him was his appreciation of the potential of the typewriter for making more than one copy at a time, although he suggested a separate ribbon for each copy rather than carbon paper, with which he must surely have been familiar. Indeed, one wonders why he did this; surely, it would have been easier to make sheets of carbon paper than lengths of ribbon.

John Pratt, a few years later, was destined to make a greater contribution to typewriter history than Dr. Francis, who seems to have limited himself to this one effort. Pratt on the other hand was not so easily deterred and his dedication was at least partially vindicated when a man called J. B. Hammond eventually mass produced one of the world's truly great typewriters based originally on Pratt's ideas. It was the only satisfaction Pratt was to get—lacking the financial resources to exploit his patents on his own, he was to see only limited reward for his efforts.

He began inauspiciously enough with the ubiquitous piano keyboard, flat paper carriage, and carbon paper. This was the prototype he had assembled in his native Alabama in 1863 before setting off for England where he protected it with a patent the following year. One significant feature of this patent, however, was to make a major impact and it was this detail which Hammond later so ably exploited. Pratt chose a type wheel as the core of the device, like so many before him, but instead of trying to move the type wheel or the carriage to produce the printed word, he devised a system based on a small hammer striking the paper against the type and this improbable combination was what ultimately ensured his place in history.

Theory is one thing but practice is something else again, as Pratt was soon to find. The trouble with his device was that it did not work, as the inventor himself was the first to admit. Back to the drawing board he went, producing radical changes in his second effort two years later. The type wheel was replaced by a conical arrangement of twenty-six type plungers, with the other features—flat paper table, carbon paper, piano keyboard—retained. Once again it did not work, but Pratt was nothing if not determined. His third effort was yet another completely different design, patented in 1866 and based upon a square index with letter selection by simultaneous depression of two keys and that, inevitably, did not work either.

So Pratt had tried a type wheel, he had tried an arrangement of type bars, and he had tried an index. He had obviously learned a great deal by the time he completed his final effort in 1866—the very year that Sholes and Glidden were beginning to modify their numbering machine on the other side of the Atlantic.

Pratt was granted a US patent for his magnum opus in 1869 (he had taken the machine back with him across the Atlantic the previous year) and he grandiosely christened it 'Pterotype.' It saw the return of the type wheel (of small diameter) and enough exciting features to make anyone sit up and listen. Pressing a knob on the right of the machine returned the carriage and wound a spring, after which typing was power assisted in a most ingenious manner. Pressing down on a key merely trips the action; the selection of the letter on a type wheel, the printing action of the hammer, and the paper advance mechanism—all were automatic. There was no space bar, as spacing was by partial depression of any key at all. This might have taken some getting used to if you were not an accomplished typist, but what a good idea!

Pratt was living and working on the other side of the Atlantic, but he leads us inevitably to his compatriot Christopher Latham Sholes, and the great Sholes and Glidden saga of the 1860s and 1870s. This was unquestionably the most important single chapter in typewriter history and one which was to revolutionize society and civilization in more ways than one. Pratt's contribution to this saga was not only practical but also inspirational, for his invention was widely reported and publicized and Sholes himself, apart from many others, is reliably reported to have been well familiar with it.

Before we actually examine this third generation mass production in further detail, however, there are still a few loose ends left over from the second generation which we need to consider. They did not change the course of history, but they were responsible for contributions of some significance.

Who were these earlier second generation manufacturers and what did they contribute? And why did new patents and inventions keep appearing in such proliferation when all the basic elements had already been devised and protected?

The first and most obvious answer is that everyone was still trying to invent machines which worked, and worked well, and that was not always easy. The second, and more important reason is that with the sole possible exception of John Jones and his Mechanical Typographer, which was wiped out by the disastrous fire,

not one of the early manufacturers was actually producing a straight typewriter. They were producing printing telegraphs, machines for the blind, and stenographic machines—and in fact they were selling them, too—but when it came to typewriters for straight correspondence, there was the inevitable resistance. People wrote letters by hand, they did not use a machine, which no doubt induced John Jones to promote his invention as a machine for the blind and not just as a straight typewriter.

Otherwise, we had Foucauld and Hughes also manufacturing machines for the blind; Brett, House, and Hughes manufacturing printing telegraphs; and Michela manufacturing a shorthand machine…and that was that for more or less a quarter of a century. It was a struggle for early typewriter inventors to overcome this universal resistance to typewritten correspondence and the struggle even continued for the better part of a generation after the machine was a reality and its future ensured.

Somewhere along the line, however, what I have simply called 'manufacture' gave way to the full scale mass production of the third generation. I do not wish to get drawn into a semantic argument as to where one stops and the other begins, common sense providing the answer, with a few cases inevitably occupying the grey area between.

Two machines dominated the beginning of the third generation…and it is difficult to imagine two more different products. One was European, the other American. The first was a spectacular piece of mechanical engineering which ultimately failed, the other started out as a woefully inadequate piece of mechanical equipment which ultimately proved a spectacular success. They were almost exact contemporaries, with the Danish contender possibly the earlier of the two by a slight margin. They certainly knew nothing of each other—not until much later—and yet they were both engaged in struggles of invention and production at the same time.

The Dane was a director of the Deaf and Dumb Institute in Copenhagen by the name of Hans Johan Rasmus Malling Hansen, whose first tentative model was from the year 1865. The design was of radial type plunger inspiration first recorded by the Italian Celestino Galli as far back as 1830 and subsequently selected by the likes of Foucauld and others. More recently, another Dane called Peters was simultaneously working on a comparable type plunger device in about 1864, but the patent protecting the invention was not filed until some six years later, by which time Malling Hansen was already in production.

Who borrowed what ideas from whom was at one time hotly disputed but is largely academic, for it was Malling Hansen who developed the principal into his famous Skrivekugle which must surely rank among the most desirable if not necessarily the rarest typewriters in the world.

The Skrivekugle (**frontispiece, 4, 58, 59, et al.**) was considerably modified in the course of its manufacture but certain elements remain constant. One of these is the essential design of the sphere, from which type plungers protrude in apparent confusion but in reality are in regular rows. These type plungers stick out like the quills of a porcupine, with a small button at the end on which the letters are engraved. The plunger in the centre of the sphere is vertical and as they spread further from the centre the plungers become progressively more angled so that they radiate perfectly from the central printing point. The whole assembly is precision engineered, and quite beautiful to behold.

This represents the common element in all models of the machine. The rest varies dramatically from one to another. Originally a cylindrical platen was attempted, enclosed in a wooden case in such a way that only the keyboard was visible from the top. Tension was supplied by a spring which required winding, while an electromagnet provided the escapement. Carbon paper was specified for printing (**58**).

58. Malling Hansen's Skrivekugle[83]

By 1870, however, the profile of the machine had undergone some dramatic cosmetic surgery. The wooden case was gone, as was the cylindrical platen. In their place was a massive steel and brass chassis weighing some 75 kilos, with the type in a sub frame above a flat paper table, with clockwork and electromagnets retained. The cylindrical platen was an optional extra. This option was also offered on later compact models featuring a curved paper frame which swung in an arc beneath the type as printing progressed. By this time, of course, the machine had reached its final perfected form, greatly reduced in size and weight and dispensing with its earlier electro-magnetic carriage release in favour of a purely mechanical system.

Just how many Writing Balls were actually manufactured may never be precisely known but certainly the number runs to many hundreds and may well have reached four figures. It was adapted for cryptography by scrambling the type to produce a simple substitution cipher which really was of no value whatsoever—not as a cipher, that is. More ingeniously, however, it was adapted for automatic telegraphy by the simple expedient of notching the type plungers so that they automatically opened and closed an electrical circuit as they were pressed down, thereby not only printing the selected letter on paper but transmitting its morse equivalent in the same operation **(59)**.

The Skrivekugle was exported to many countries, from Egypt to Peru, and was manufactured under license overseas and enthusiastically received at Exhibitions all over the world. It still retains this appeal, today remaining one of the most desirable of all historical typewriters.

The Skrivekugle shares this distinction with its American contemporary, which Sholes and his friends were in the process of developing at much the same time as Malling Hansen was in the final stages of assembling his first wooden cased model.

Sholes may, perhaps uncharitably, be said to be the most famous of all the typewriter inventors never to have invented anything, yet his was the machine which virtually began the modern typewriter revolution single-handedly. Such is the irony of typewriter history.

The Sholes adventure began in the mid-1860s in a small workshop in Milwaukee which belonged to a man called Kleinsteuber. All the local amateur inventors used to gather there, each working on his own project. Sholes, who was a printer by trade, was trying to make a machine which would automatically number pages. A friend called Samuel Soulé was helping him. Carlos Glidden was working on a mechanical digger. Henry Roby, Sholes' neighbour, was making a magician's clock.

It seems that one day, Glidden arrived with an article from the *Scientific American* about Pratt's typewriter invention, and he suggested to Sholes that he modify his numbering machine into one which printed letters.

No problem to the resourceful Sholes, who soon produced a device with a single telegraph key controlling a type bar which printed by up stroke on paper held beneath a flat plate. Easy—no trouble at all! A piece of carbon inserted between the various elements, in the age-old manner, sufficed for printing.

As the group of amateur inventors gathered around to admire this primitive device, it all seemed so easy that they were amazed no one had done it before. Enthusiasm ran high and they all pitched in with advice and suggestions. One of Kleinsteuber's machinists called Schwalbach gave a hand, as did Roby and the others. But it was mainly Glidden and Soulé who provided the help and inspiration which Sholes needed to advance to the next stage, which was to turn the single little telegraph key into a fully fledged typewriter with a separate key for each letter.

The date was September 1867, by which time rival typewriter inventors had come up with some very sophisticated machines indeed, making the amateurish efforts of the Kleinsteuber lads seem risible and more than a little primitive, despite their optimism and enthusiasm. What they did was to cut a circle in the top of an old

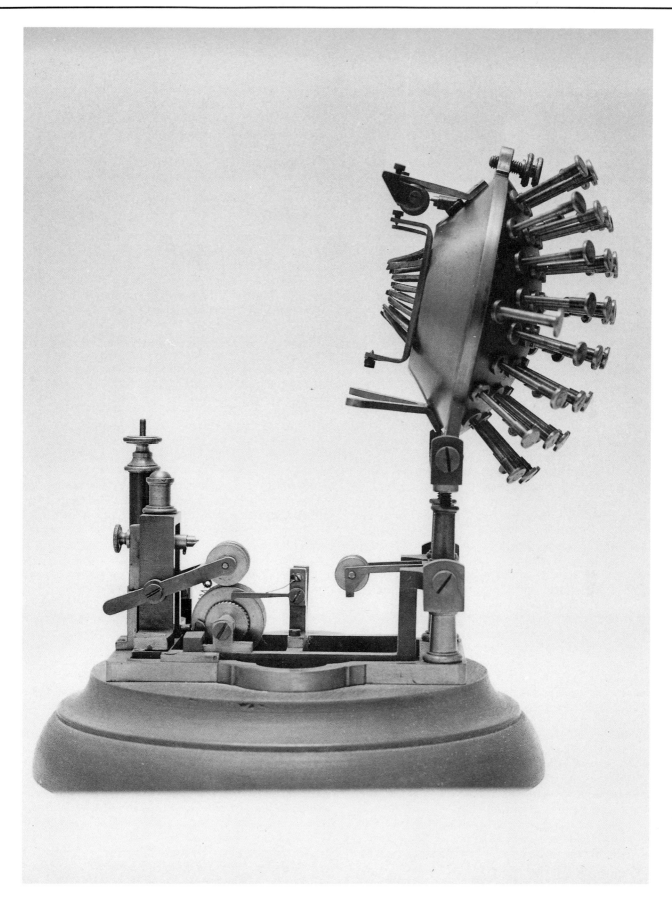

59. Malling Hansen's Skrivekugle (Courtesy of Arne Magnussen)

kitchen table, hang a ring of type bars into it, and by means of wires down one side of the machine and up the other, connect the type bars up to a primitive lateral keyboard composed of protruding levers. The paper was in a flat frame above the type basket, with a ribbon above the paper providing the impression. The flat paper frame had a weight-driven escapement controlled by a cord which tended to snap during typing, allegedly with alarming consequences.

There was absolutely nothing original in any of this, as is immediately obvious, but then we have come to realise that all the individual elements had already been invented decades before. It did not seem to matter to anyone, however: Sholes, Glidden, and Soulé immediately protected their invention with a patent application and, in a rush of blood, turned their efforts to manufacture. The trouble was that they had no money or experience, and they were already having some serious difficulties even covering Kleinsteuber's modest bills because in order to meet their current commitments, to say nothing of their long term plans, they had been obliged to make a number of models of the machine. One of these models went to the patent office, while others were used for promotional purposes to try and attract an investor. Some were even dropped off with friends to be tested to destruction…and the tests apparently achieved their purpose with disconcerting speed.

Charles Weller, a telegraphist, received one and a shorthand reporter called James Ogilvie Clephane had another. Despite the fact that the inventors had now abandoned the use of kitchen tables and built the frame and the machine as a unit, the condensed version sported improvements which were essentially cosmetic and little else. It was still a big and clumsy brute, using a flat paper table and a weight driven escapement with a mass of ugly key levers, connecting wires which snapped and type bars which stuck and jammed and cords which broke and weights which dropped.

At this point, a three hundred pound bearded giant of a man called James Densmore paid, sight unseen, the most handsome sum of six hundred of his hard-earned bucks for a quarter share in this contraption. Of all the potential investors approached by Sholes, Densmore was the only one who had taken up the share offer, even though his own solvency at that stage was already in some doubt. Perhaps this is what induced Densmore to take such an active part in what was initially intended to be little more than a sleeping partnership. For when he eventually saw what he was meant to be getting for his $600 investment, not to mention his rash commitment to finance the manufacture of the machine, he realised that unless he could conjure up something dramatic, his money was as good as down the drain.

It was now March 1868. Densmore demanded action, and he was the kind of man who had a way of getting what he wanted. Two months later he was able to file a patent for a greatly improved device which was smaller, more compact, sported a piano keyboard and was contained within a wooden case—altogether more pleasing, despite the flat paper table. High on optimism and enthusiasm, Densmore sank a further grand into a premature venture at manufacturing the instrument, aborted after a mere fifteen unsatisfactory units had been produced.

Sholes, who some years earlier had confidently declared that he '*had the right idea at last*,' was by now just as confident that he was never likely to get the cursed machine right, however hard he tried. He was fed up with the whole thing and particularly with having Densmore breathing down his neck, for by now the man had bought out Soulé's and Glidden's shares and his financial commitment was multiplying at an alarming rate. Sholes wanted out, but the one thing Densmore did not want was to own the whole venture on his own and the one partner he could not do without was Sholes himself, so the poor man's entreaties were to no avail.

Out came one prototype after another, as if Sholes realised that the only way he was going to get Densmore off his back was by succeeding. But success was not even in sight as yet. By 1869 and 1870, changes were indeed being wrought and a recognisable typewriter was beginning to emerge but the prototypes were being thrown out by Densmore almost as fast as Sholes could produce them. They simply did not work, he declared—probably loudly and not at all kindly, for he is said to have had a booming voice and an uncouth personality.

A measure of Densmore's own desperation may be gauged from the fact that he tried to sell out the whole business to the American Telegraph Company—and almost succeeded, too, except that a young employee of that company threw a wrench in the works by convincing his bosses that he could build a better writing machine cheaper. In the event, he failed to produce a competitive typewriter at the time, but he did do so some years later. His name—Thomas Alva Edison.

What Edison did on this first occasion, as he did so often before and after, was to design and patent a device using well established principles long since invented by others. In this instance, it was the comb and pin-barrel idea with a constantly rotating type wheel and an electro-magnetically operated hammer for the impression. He later used the hammer again, in a different way, on his Mimeograph typewriter **(17)**, but that is another story. History is more likely to remember his printing telegraph for dashing Densmore's dreams rather than for its own intrinsic merits.

Yet another interesting detail emerged from the event which gives us some insight into the sort of cross-pollination which was taking place. Edison's telegraph used a type wheel which travelled along the line in a manner comparable to modern electric golf ball machines. Once again, he did not invent this principle, but was astute enough to recognise its potential. What it meant was that a continuous roll of paper could be used and the telegrams torn off at the end of each printing. Sholes at this time was using a different arrangement altogether, with the platen set at right angles to the way it now is on modern machines, rotating space by space as each letter was typed and moving along its axis the required amount at the end of each line for the next line to begin. Edison later claimed it was he who introduced Sholes to the idea of the modern platen arrangement, and on this occasion there may be some truth to the claim, for shortly after Densmore's abortive commercial excursion into telegraphy, Sholes began to produce prototypes with modern platens. Another decisive step had been taken.

Of course, modern platen or no, the machine remained an up stroke type bar instrument and produced its typing beneath the platen, so that the text was not immediately visible to the typist. This was a curse but Sholes was stuck with it because there was no other way that his various elements could be combined. So long as it remained up stroke with the basket of type bars beneath the platen, the typing would be blind and the best that could be done was to put a brave face on it and pretend it did not matter. In later years, Remington mass produced this basic design for three decades, claiming that a typist no more needed to see what was being typed than a pianist needed to see what keys had been struck. They simply ignored the obvious fact that the pianist, unlike the typist, could immediately hear a wrong note whereas the typist continued blissfully unaware of the mistake, usually until the end of the page.

Wrong keys were not the only curse of these blind machines either—early ribbons had a recurrent habit of jamming or coming adrift from their spools, and it was only at the end of the page that the resultant 'stencil effect' became apparent.

However, Sholes and Densmore were fully committed to the up stroke type bar design and struggled to iron out the bugs before total financial disaster overwhelmed them. By the time Densmore had made yet another abortive attempt at manufacture, he was out of pocket to the tune of more than $13,000, which in those days was a tidy enough sum even for a man who was not virtually broke to begin with. Unable to sell the venture outright, he began to sell off bits to anyone who would buy a stake, notably to his brothers Amos and Emmett. The confusion which surrounds the precise ownership of the invention at this stage is not just the result of historical neglect; such was the complicated buying and selling of tenths and fractions of tenths that in all probability not even Densmore himself was altogether sure who owned what. The only thing he was sure about was that Sholes had to keep a share and had to stay chained to the workbench until they succeeded.

And succeed, finally, is what they all did, largely through the offices of a fast-talking gentleman with the unlikely name of George Washington Newton Yost.

By late 1872 or early 1873, Densmore had got nowhere—the Sholes prototypes, albeit vastly improved, still could not be made to work properly. Calling in Yost at this stage must have been a final act of a truly desperate man, fuelled by the probable conviction that they were not going to make it without help, however much it cost them. Yost was no philanthropist and the precise price tag for his involvement is uncertain. Given Densmore's intense reluctance to part with more money and Sholes' inability to do so, Yost's price was probably a share of the action because he was to turn up in later years as a fully fledged partner in early typewriter retailing, and went on to forge an illustrious career for himself with the interesting machine to which he later gave his name **(60)**.

60. Yost No. 4 (Courtesy of This Olde Office, Cathedral City, California)

What seems certain is that it was Yost who, recognising the potential of the invention, turned his attention to launching it commercially not by embarking on yet another hopeless manufacturing effort of his own but by selling the idea to those he considered the most likely potential clients: Remington.

It was to prove an auspicious stroke of commercial astuteness on Yost's part. Old Eliphalet Remington had built up a thriving gun making business which virtually spanned the 19th century, reaching its peak during the Civil War. Of course, peace is no boon or blessing to a gunsmith and the original Remington's sons Philo, Samuel, and Eliphalet Jr., who controlled the business after their father's death, were quick to realise that they had to diversify their industrial manufacturing potential into peace-time pursuits. They were already making treadle sewing machines as part of this policy when a letter arrived from Densmore extoling the virtues of Sholes' machine to replace the pen by writing through the manipulation of keys, and asking whether Remington would be interested in seeing the article with a view to manufacturing it.

Philo Remington called through to Henry Harper Benedict, one of his directors, in an adjoining office. Benedict's principle area of responsibility was for the sewing machine division, of which he was treasurer. The two men briefly discussed the potential of such a writing contraption and decided to take a look at it. Densmore and Yost made the journey with their latest model early in 1873.

Sholes himself, who would hardly have been an asset at such an encounter, was wisely if uncharitably left at home.

The meeting was held in a local hotel. Present, along with Densmore, Yost, Remington, and Benedict were two technicians from the Remington works, Jefferson M. Clough and William K. Jenne. They were all together for a couple of hours and then the meeting was adjourned to allow the Remington delegation to confer. Yost had done all the talking while Densmore stayed well in the background, trying hard to remain unnoticed and inconspicuous (which did not come naturally to him) and determined to ruffle as few feathers as possible. This was by no means easy for him, and, if eye-witness reports are to be believed, one can only marvel that Yost even allowed him in the room at all.

Henry Roby's narrative goes some way towards explaining why.[75] *'Densmore's large, cumbersome body and hairy, red face seemed in harmony with his irritable manner and temper...'* was how Roby described him, adding that he was

> *...a great ponderous beefy-looking man of nearly three hundred pounds weight, with a florid complexion, a great shock of red hair, a shaggy beard, the eye of a hypnotist and the heavy jaw and animal force of the great Hyrcanian bull in Quo Vadis...His attire was characteristic of his remarkable personality; a seedy, crumpled hat, a long shabby coat, a vest off color and swearing at everything*

else about him, trousers too short by some inches for his short loggy legs and coarse old woolen (sic) socks thrust into his low cut shoes. His shirt, that was sometimes washed and sometimes not, was seldom adorned with either collar or cravat...

One suspects that Yost may well have succeeded in cleaning Densmore up more than somewhat for this vital meeting, but just as the early development of the Sholes and Glidden prototype was a personal triumph for Densmore, the successful outcome of the Remington negotiations was a personal triumph for Yost. And successful it was, for the participants entered into a tentative agreement that same day and subsequently hammered out the details of a manufacturing contract which was eventually signed on March 1st 1873.

Remington, however, was not prepared to take too many chances, if the terms of the agreement were anything to go by. All they were actually committed to doing was manufacturing the instrument on behalf of Sholes, Densmore et al. It is true that they had to invest a considerable amount of time, effort, and expertise in the venture in order to make a marketable product out of the model they were given, but they were not inclined to be philanthropic about it. They undertook to manufacture an initial batch of 1,000 machines with an option for a further 24,000—a wild projection of almost manic optimism at the time, never to be realised. (In the event, only some 5,000 were actually made before the model was changed and the early design was dropped.) In return Remington received an advance of $10,000, with more to come,

from Densmore and Yost who formed a partnership to introduce the product into the retail market.

Those must have been hectic days indeed for everyone involved in the undertaking, and the number of claimants for handouts seems to have grown all the time. Sholes and Glidden had been involved from the start, and of course were now entitled to royalties. There was Densmore, who was most out of pocket, and there was Yost, and then Densmore's brothers Amos and Emmett, who had bought shares in the venture when it seemed little more than a pipe dream and who were still actively promoting it. On the 4th June, 1873, for instance, Emmett Densmore wrote a letter to Queen Victoria inviting her to inspect the machine—a little prematurely, in the event, since manufacture did not actually begin until September of that year—but Her Britannic Majesty was apparently not amused, for on the obverse of the letter, which has survived, was the single word, presumably written by the Queen herself, *'Declined.'*

What it proves is that even though the machine itself was now physically out of their hands, the business side of the venture remained active enough and this was destined to become progressively more complex in the years ahead. Yost was everywhere. He and Densmore set in motion the first sales organization in anticipation of the delivery of the machines from Remington early in 1874. One of the first machines off the production line, previously in the author's collection, was serial no. 33 **(61)**, which remains the oldest surviving example in private hands.

61. Sholes and Glidden Type Writer

Meanwhile it had taken Remington the better part of a year to make a practical instrument out of the Sholes and Glidden prototype. This task had been entrusted to the two technicians William Jenne and Jefferson Clough. Both these men had been active in Remington sewing machine **(62)** manufacture and Jenne in particular was responsible for redesigning the typewriter by placing it on a sewing machine stand, with a pedal for carriage return. Another Remington employee involved at that stage was a mathematician by the name of Byron A. Brooks, who later went into typewriter manufacture on his own with a machine to which he gave his name. It is indeed fascinating to follow the many threads of people's lives through these early typewriter years—the same names recur time and again.

Remington had figured that it made good commercial sense to hang the typewriter on the sewing machine peg, so to speak, quite apart from the obvious technical prejudices of Jenne and Clough. The sewing machine was after all a well known and well established domestic appliance which could only help to familiarize the public with the new-fangled invention. '*The Type Writer in size and appearance somewhat resembles the Family Sewing Machine*' was how it was initially promoted.

Meanwhile the manufacturing side of the story is almost as complex as the retail side. As they began to wrestle with the technical complexities of the task, Remington had to make a few deals of their own and the bill was invariably passed on to Densmore and company. Jenne and Clough demanded, and received, a flat fee per machine for the improvements they had made. Then it was found that a certain Charles Washburn had already patented the carriage design which they adopted, so they had royalties to pay to him as well. On top of that there were both Sholes and Glidden to consider, since they held the original patents. The machine so far was upper case only; when Remington began thinking of introducing the upper and lower case model it was soon discovered that Byron Brooks already held a patent for that particular development.

The one thing they did not need to worry about was the keyboard. Every major invention had a keyboard or index of some form or other, so there were no royalties to be paid on keyboards. Type bars, up stroke action, platens, ribbons, circular keys, keyboards: none of these was original to the Sholes and Glidden, and yet one of its design features was unique and it has remained with us to the present day. This is the order in which the letters appear on its keyboard **(51)**.

62. Remington Sewing Machine

63. Blickensderfer 5

QWERTYUIOP…we know it all too well. Although we touched upon this subject briefly in the previous chapter, it is worthy of a little further analysis.

Why did they do it? Men and women have been asking themselves that for the past hundred years and coming up with a number of intriguing answers…all of them wrong! No other aspect of the typewriter has been subject to such intense scrutiny and analysis because time and again attempts have been made to correct its obvious deficiencies. But to no avail. Our existing arrangement—commonly known as the Universal keyboard—has only ever had one significant alternative. This was the not inappropriately named the Ideal keyboard, which was at one time quite widely available on machines such as the popular Blickensderfer (**63, et al.**) and Hammond (**64, et al.**). However, even these makers were eventually obliged to offer the Universal as optional, and ultimately the Ideal was dropped altogether.

It all began with Sholes and Glidden, then, and a close look at machine number 33 provides some interesting facts. For the sake of reference, the keyboard arrangement of this particular instrument is as follows :

23456789-,£
QWERTYUIOP:
"ASDFGHJKLM
"ZCXVBN?;.!

Not all that different then, from the present-day sequence, nor from that on the accompanying illustration **(51)** from a contemporary Remington leaflet which, while it has slight differences in detail from the above order, nevertheless has a familiar look to it.

Now these keys, as is known, controlled type bars in a circle suspended beneath the platen. Beginning arbitrarily with the same *2*, for no better reason than that it is the first letter on the keyboard, and proceeding in a clockwise manner around the circle (the *!* at the beginning of the second line in the schematic representation below following immediately after the *:* at the end of the first, so that the final *"* and the initial *2* finish up being adjacent to each other) the order of the type bars in the circular type basket is as follows :

2Q3W4E5R6T7Y8U9I-O,P£:
!M.L;K?JNHBGVFXDCSZA" "

What is the explanation for the legacy which Sholes (and Glidden) left to mankind, possibly for all eternity, in his sequence of the keys on his keyboard?

Several alternatives have been suggested. The first is that the order was determined by type bar clash. The proponents of this theory claim that the somewhat crude and sluggish action of the early typewriters meant that even typing with only two or at most four fingers, in the days before touch typing was introduced, a type

64. Hammond (Courtesy of CNAM, Paris)

bar did not always have time to drop to its position of rest before colliding with another type bar on its way up to the platen. In an attempt at minimizing this phenomenon, we are told, Sholes contrived to place the type bars of the most frequently used letters as far away from each other—ideally diametrically opposite each other—as possible. He then simply connected the type bars to the most convenient key on the keyboard.

This sounds both plausible and commendable. Sholes, after all, was a printer by trade and was therefore familiar not just with the practise but also with the theory of letter frequency in the English language. *'The tendency of type bars on all the Sholes models was to collide and stick fast at the printing point, and it would have been natural for Sholes to resort of any arrangement of the letters which would tend to diminish this trouble.'*[41]

However likely this may appear, it is quite simply nonsense. The most widely accepted order of the frequency of letters in the English language is as follows:

E T A O N I R S H D L U C
M P F Y W G B V J K Q X Z

There are minor disagreements among the experts—some invert the positions of N and I for instance—but there is no dispute at all that the letters E T A O I N S H R D L U represent the twelve most commonly used letters in the language. Sholes would have known this. He would also have known that ER and RE are the two most frequently used digraphs. Even the most superficial glance at the type bar arrangement, therefore, will suffice to explode the theory of letter frequency. If it had indeed been Sholes' intention to place the most frequently used letters as far apart as possible, he should have positioned E and T diametrically opposite each other, not virtually adjacent to each other, with E and R as far apart as possible as well.

So it was not letter frequency which determined the arrangement, and it obviously had nothing to do with the ease of operation of the keyboard either. Only one or two fingers were used for typing at the time, and were ease of operation the determining factor, one might have expected to find the most frequently used letters clustered around the centre of the keyboard with the others radiating outwards in direct relationship to the infrequency of their potential use. Once again, this is clearly not the case.

Another theory, even more speculative than the previous ones but at least more difficult to dismiss, stems from the fact that, early on, someone realised that the letters of the word TYPEWRITER are all located on the same line. This was done intentionally, it was said, in order to allow salesmen to type the word quickly and thus impress potential clients with their speed.

Reality, however, does not appear to substantiate this implausible claim because the letters of the word are so badly jumbled. If producing the word TYPEWRITER, at speed, is of major importance and the letters need to be scrambled to conceal the deceit, then surely the sensible approach would have been to arrange the letters so that each hand typed an alternate one. On one side of centre the letters T P W, with Y E R to the other side, for example. Even more to the point, of course, is that your potential buyer might well require more of a demonstration than the typing of a single word, so you might find yourself in trouble all over again.

No, this is not plausible, even though it might have sufficed for Mark Twain who was one of the Type Writer's first clients. He allegedly demanded a 60 second demonstration of typing skill prior to purchase, and was quite dazzled by the speed of the operator (or 'type writer,' as she was then called) only to discover on later examination that she had simply typed the same memorized phrase over and over again.

What other explanations are offered? Roby, who with Oden was responsible for the perpetuation of so many myths surrounding the early days of the Type Writer,[72] was one of those who claimed that the keyboard arrangement was the product of Sholes' being a printer by trade and thus naturally opting to place the letters in the same sequence as they appeared in his trays of type.[75] '*I'll arrange the letters in the same order as that of a printer's font,*' Sholes is supposed to have said, '*and then they will keep out of each others' way.*'

This is not a bad theory, either, because in addition to solving type bar clash, a printer would have memorized the order in his trays at an early age and selecting the correct letters would become second nature to him. All perfectly plausible—except that the order does not conform to that of a printer's trays at all, any more than it conforms to frequency of use. And even if it had conformed to a printer's trays, it would not have kept the type bars out of each others' way.

So we are left with the conclusion that there may well be no legitimate reason whatsoever for the QWERTY arrangement. There are still traces of an alphabetical order—possibly it began alphabetically, but eventually it seems likely that most of the letters were simply arranged at random in the order they were extracted from a box of bits as the prototype was being assembled, perhaps. It did not matter at all to the inventors when there were so many more important things to worry about, such as whether the project would even succeed in getting off the ground. Money was the worry in those days, not the speed of typing.

Indeed, even a decade or more after the typewriter had been commercially launched, operating speed was still of no significance whatsoever. The most commonly used method of typing in those early days was the 'Columbus System,' so named because a discovery was involved every time a key was struck, and this system required the use of only two or at most four fingers and indeed, typists were actively discouraged from using more. Mrs. L. V. Longley of Longley's Shorthand and Typing Institute in Cincin-

65. Caligraph 2 (Courtesy of Bernard Williams)

nati, Ohio, was the first person to go on record to claim that ten fingers should be used for typing and that was as late as 1882. Five years after that, a trade magazine called *Cosmopolitan Shorthander* was still editorializing that *'the best operators use only the first two fingers of each hand, and it is questionable whether a higher speed can be attained by the use of three.'* The issue was not effectively resolved until as late as 1888 and even then it must have taken years for the idea of ten-finger typing to filter through the system.

1888 was the year when a historic challenge was held between the world's two top-ranked typing heavyweights. Frank E. McGurrin of Salt Lake City, the father of touch-typing using all ten fingers on a Remington Model 2 versus a certain Louis Taub of Cincinnati, using four fingers on a full keyboard Caligraph **(65)**. McGurrin's convincing victory heralded the beginning of acceptance of ten finger typing and the superiority of touch over sight (at least as far as typing is concerned), even if it did not result in any immediate changes of attitude.

further 1,300 or so arrived during the second year and another 1,500 more the year after. Manufacture was now well into its stride, but the retail side was letting them down badly. Of course Remingtons still had the market, such as it was, strictly to themselves and their only possible competitor was the Hansen Writing Ball, which made no commercial impact whatsoever in the United States. The trouble was not competition but the fact that there was no market, or not much of one anyway. Of course, it might convincingly (if uncharitably) be argued that the machine itself was at least partly to blame for its own marketing problems, as witness some contemporary correspondence which accompanied the (*soi-disant*) Perfected version of an example of the machine **(209)** when the author acquired it some years ago. The frustrated efforts of '*Rose?ary ?oore*' to produce a piece of acceptable copy is reproduced without further comment **(210)**, as is Remington Typewriter Company's reply which concedes, with almost mind-numbing under-statement, that *'...you require some adjustment made to your*

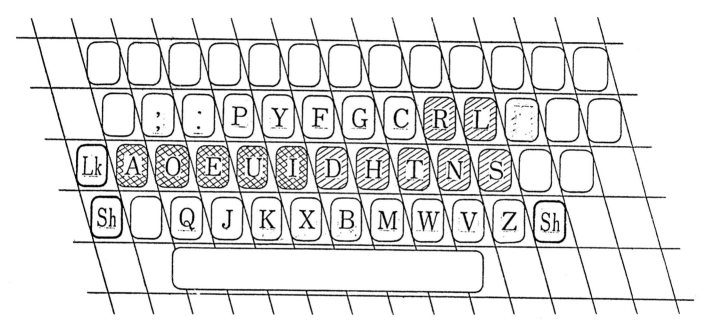

66. Dvorak's modified keyboard[94]

For the moment at least, enough said of the QWERTY keyboard which has so profoundly affected all our lives. Many attempts have been made in the course of the past century to modify it and improve on it but without success. For all its inherent faults, it is likely to remain with us forever. A man called Dvorak made one of the most serious of many studies of the subject well over half a century ago, and eventually proposed his own modified and vastly improved keyboard **(66)** as an alternative, the twelve most frequently used keys in the accompanying illustrations better located and divided between right and left hand.[94] As a result, the number of awkward strokes on a standard keyboard **(67)** and those on his own modified keyboard **(68)** are reduced by a factor of ten, the height of each crossbar joining two keys in the above illustrations indicating the frequency of awkward strokes between those keys on both alternatives.[94] Alas, to no avail.

Back in 1874 however, all this was an utter and risible irrelevance—simply launching the new invention onto the market was all that mattered. Remington was employing some thirty people on it at the time, manufacturing and assembling some 550 Type Writers during the first year. It does not sound like much these days, but it was more than they could find homes for.

Just how many they did manage to sell is not known, but we do know that they still had a bunch of them on their hands when a

typewriter.' **(211)** True, the machine itself was a couple of decades old by then, but contemporary records reveal that it was unlikely to have performed better even when it was brand new.

Remington had other troubles too. For a start, the horse trading that was going on in Type Writer Company stock between the original shareholders and Yost and the three Densmores shows just how unstable the partnerships were at that time. This instability manifested itself in ever changing combinations of retail agents. Yost and Luxton, Densmore Yost and Co., Locke, Yost and Bates, Fairbanks and Company…all of them tried over the years to make a go of it but with limited success.

The *Locke* in Locke, Yost and Bates was a man called D. R. Locke, a famous humourist at the time, better known as 'Petroleum V. Nasby.' In the autumn of 1874, he and Mark Twain first saw a Type Writer in a shop window. It obviously made a lasting impression on them both—Mark Twain bought the machine on the spot, while Locke bought a share of the retailing organization, or *'general agents'* as they liked to call themselves.

The firm of Locke, Yost and Bates briefly held the franchise before relinquishing control to Fairbanks and Company, the well established scales manufacturers who were destined to be the last of the unsuccessful agents of the Type Writer. A further 2,000 or so machines had been manufactured by then, and sales were still dis-

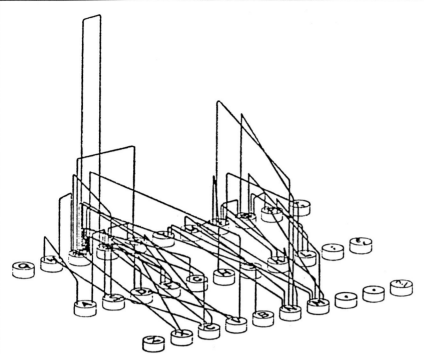

67. Awkward fingering on a standard keyboard[94]

mal. It was an uphill battle for Fairbanks, who seem to have done little right as agents, apart from momentously appointing a man called Clarence Walker Seamans as manager of their typewriter division.

What an incestuous little world it was, this early typewriter microcosm! The same people pop up time and again. This man Seamans, who now managed the typewriter division of Fairbanks and Company, was the son of the man in charge of gunsmithing at the Remington factory. Young Seamans had worked there himself as a lad, helping his father. When the Fairbanks outfit decided to take on the typewriter franchise, Yost himself recommended that Seamans be put in charge of the agency. Philo Remington opposed it on the grounds that Seamans was too young but Remington's co-director Henry Benedict supported the appointment.

Fairbanks and Company, with Seamans in charge of the typewriter division, held the agency from 1878 to 1881 but fared little better than their predecessors. Sales in that final year were still a mere 1,200 units and this goes some way towards vindicating the infant typewriter as the cause of its own problems, however poorly it performed. The wounds in those early days were not all self-inflicted. Even if the machine had performed better, sales were unlikely to have improved, because by 1878 Remington had already totally redesigned its product into an open frame, upper and lower case Model 2, soon to blossom into the ubiquitous Model 7 **(220)**, the essential features of the old Model 2 remaining in use through one minor model change after the other until as late as 1908. Even so, it was years after its 1878 introduction before its sales vindicated it as the most successful design of its time.

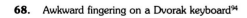

68. Awkward fingering on a Dvorak keyboard[94]

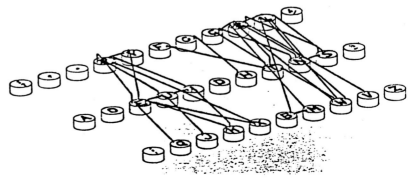

All of which was not immediately apparent back in 1881 when Fairbanks and Remington parted company, with the factory taking over the retail organization themselves and putting Seamans in charge. The arrangement lasted barely a year. Negotiations resulted in Seamans breaking away from the Remington organization and teaming up with our old friend Benedict and a newcomer called William Ozmun Wyckoff. Newcomer? Did we say 'newcomer'? Wyckoff had been Yost's star salesman back in the days when Yost had been involved in early efforts to retail the Sholes and Glidden. Although it took many years for sales to pick up, as we have seen, and Yost's efforts at that particular time had been dismally unsuc-

cessful, Wyckoff had gained invaluable training, although his credentials went back even further than that. For when he first saw the Type Writer, back in 1874, he had tied up the selling rights for Central New York State, and had had the conviction and prescience to place one of the new machines in his own offices.

Wyckoff, Seamans and Benedict teamed up in 1882. Theirs turned out to be a partnership made in heaven, and one which was destined to provide one of the pivotal points in modern typewriter history, for it may be said quite definitively that the exponential growth of the typewriter market in the latter part of the 19th century was due in large part to their dynamic involvement.

69. Crandall New Model

Chapter Four—
Third Generation: The
Age of Mass Production,
Part 2

Once again, having partially dispensed with chronology, we must pick up some threads from previous years, because the most significant detail to emerge is that by the time the Remington was finally successful, it no longer had the field to itself.

It may quite rightly be argued that Remington never really had the field to itself in the first place, because a number of typewriters had already been manufactured long before, and one of these, at least, had been a direct competitor. Machines for the blind and printing telegraphs were sufficiently different and specialized to make direct parallels unsatisfactory. But the Hansen Writing Ball—the *Skrivekugle*—was a different matter; not only was it a better machine than the Sholes and Glidden but it was also manufactured and sold some considerable time earlier and its sales continued well after the American machine had already been successfully launched.

While all this is quite true, however, the production of the Writing Ball (**frontispiece, 4 et al.**) was always limited and its sales were small. It was never aimed at the mass market and certainly not in the United States. In that respect the Sholes and Glidden (**1, 12 et al.**) did indeed enjoy a virtual monopoly, such as it was, during those unhappy early years.

However by 1878, when the Sholes and Glidden had become the Remington Model 2, its monopoly had all but come to an end. The following year, another American by the name of Lucien Stephen Crandall had put the finishing touches to a fine type sleeve

machine for which he was granted a British patent in 1879 and the US patent two years later, by which time his machine was already on the market. This was not Crandall's first contribution to our history. In 1875 he had patented a curious device with eight keys controlling eight type bars on each of which were six characters. Predictably enough this design remained firmly on the drawing board, and Crandall may be assumed to have suffered all the agonies of jamming type bars and faulty alignment which all contemporary inventors of type bar devices were experiencing. His next machine therefore, which he not only patented but actually manufactured, abandoned type bars in favour of a long type sleeve controlled by a straight three row keyboard. The machine was not a hit by any means, but it launched Crandall on a long and moderately successful career which spanned some thirty years of versatile inventiveness.

That first machine which Crandall marketed was replaced in 1886 by an improved version, appropriately labelled New Model. This is the design for which he is best remembered by present day enthusiasts, and the accompanying illustration (**69**) of the example formerly in the author's collection is lasting testimony to its enduring eloquence. Gone was the straight keyboard in favour of a curved two row design of elegant profile, but the type sleeve concept was retained. It was still used when the design underwent a further modification with a return to the straight three row keyboard on the Universal Crandall in 1893.

Two years later he patented a different typewriter altogether which he manufactured in reduced numbers under the name Improved Crandall. He returned to his initial inspiration from twenty years earlier on this 'improved' machine, with its strange design using twelve horizontal radiating type bars with six characters on each bar and a straight three row keyboard. The curious feature of this invention was that the type bars did not actually strike the paper on their own, but merely moved into the required position whereupon they were hit by a small hammer to produce the impression. It is easy to see what Crandall was driving at, but it was a sterile idea and such anachronistic efforts merely demonstrate that an inventive mind had run its course to exhaustion.

Meanwhile he had also been involved in other ventures of lesser importance. One of these involved a double keyboard up stroke machine which he briefly called American Standard and which landed him in court with Remington who successfully protested against his use of the word Standard in the name of the machine—the design eventually became the Jewett, under which title it is best remembered. Then there was his up stroke type bar device with a three row keyboard which he patented in 1886 and manufactured from 1889 in several modified versions under the name International (**180**).

70. Gardner (Courtesy of Christie's South Kensington, London)

More speculative however are contemporary reports linking Crandall to yet another machine called Victoria. This is one of the names under which a curious device, better known as the Gardner **(70)**, was marketed. Crandall is reported to have been involved in manufacturing a machine bearing the name Victoria and sporting unusual features of undeniable Crandall inspiration, not least of which is its vertical type sleeve with printing by means of a hammer. However, precisely what part Crandall played in this design, if indeed he was involved at all, has not been established.

Now that former Crandall machine—the one which used type bars and a hammer—is an interesting hybrid because it leads us neatly to a man called James Bartlett Hammond, a contemporary and competitor of Crandall's. He turned the hammer principle into one of the best and most successful machines ever produced **(71 et al.)** and this may well have induced Crandall to try something along similar lines. The two men knew each other from the earliest days and had in fact locked horns over the use of Pratt's type wheel patent, which Hammond claimed Crandall had infringed. It was Hammond's victory in the ensuing legal battle which forced Crandall to elongate the type wheel into a sleeve, or cylinder.

Pratt, it will be recalled, had spent years trying to perfect his writing machine but he was never able to proceed beyond the patent office. He returned to the United States from Great Britain in 1868 and some years later struck a deal with Hammond permitting him to use the Pterotype's type wheel, having failed to raise the capital required for manufacture himself. During this period he became aware of Sholes' efforts and Remington's involvement with them, and he is reported to have approached them himself with a view to promoting his type wheel design. Nothing came of this as Remington was committed to Sholes' type bars, and Hammond was equally adamant about the advantages of his own emerging ideas.

71. Hammond Type Writer

What eventually evolved was a different design altogether. Hammond was not happy with the type wheel. There was the problem of inertia and the complication of the wheel's having to turn a different number of degrees for each letter. His eventual solution was to use an arc of the circle only, at first in two narrow swinging sectors and then in a single sector fitted to a relatively heavy wheel. Printing was performed by means of a hammer striking the paper against the type face from the rear, cushioning the blow by means of a narrow strip of rubber, cloth, or leather. A curved two row keyboard with solid wooden keys of rectangular section graced the front of the instrument (71). On later models these solid keys were first exposed (50) and eventually replaced by thin key tops (81).

It was a great design but there were still problems to which Hammond could find no satisfactory solution. With the shuttle at the front and the hammer at the back, on the same horizontal plane, there was no way for the sheet of paper to enter and exit from the machine in the conventional manner. The best that Hammond could devise was a carriage containing a cylindrical basket into which a rolled up blank sheet of paper was first inserted so that the leading edge, between feed rollers, was on its way out. This was a clumsy enough arrangement to have doomed a lesser machine to the scrap heap and it is a lasting tribute to the greatness of the Hammond in every other department that this inconvenience was tolerated at all, let alone dismissed, by its enthusiastic following.

The Hammond was indisputably the best machine of its time. The regularity of the typing was unparalleled and the ease with which type shuttles could be interchanged gave it a versatility unequalled in any other machine, with the possible exception of the Blickensderfer. Its manufacture in a wide variety of models, all using the same basic system, continued until modern times—the Hammond itself continued until the 1920s, but the VariTyper version is still in use to this day.

Back in 1881 when the Hammond was in its infancy, however, things were very different indeed. Everyone was still struggling against an apparently uncompromising public. Remington, having had the market virtually to itself at the start, was only just beginning to reach world-wide annual sales in the four figures. They were still bailing themselves out of the trouble they had experienced with the earlier treadle machines and no sooner were they achieving some recognition than they were having to face competition from Crandall and Hammond.

And there were others, too, who were demanding their share of the tiny market. One of these was a man called Thomas Hall of New York whose credentials, like those of Pratt, could be traced back to the 1860s. Hall spent the first part of his inventive career working on a type bar machine of down stroke design which he eventually patented in 1867. He positioned the keyboard at the top of the machine, with the circle of type bars beneath. The type bars struck down on a flat sheet of paper inserted into the bottom of the machine underneath the type

basket. With such additional features as a ribbon and differential spacing, the design might have proved successful, for Hall is reported even to have received advance orders for the device, but manufacture was aborted due to internal disagreements between Hall and his partners.

It was perhaps still a few years too early. Just what Hall did in the interim is not documented, but fourteen years after his first patent he was back in business with a machine (5) totally different in design not only to that of his earlier effort but also to those of his competitors. There was the Malling Hansen type plunger, the Remington type bar, the Crandall type sleeve, the Hammond type shuttle and into this confusion came Hall with an indicator machine using a square rubber index! What were potential typewriter clients, as were to be found in those days, to make of the conflicting claims of these rival manufacturers?

Hall looked carefully at the technical and mechanical problems his competitors were experiencing and came up with a small, compact portable machine in an elegant rectangular wooden box. No type bars, no type wheels, no ribbons, no problems of alignment or inertia, no aching inky fingers pounding heavy keyboards, unjamming ribbons and type bars, or rolling up sheets of paper and inserting them into baskets. Hall's solution was a small square rubber index attached to an indicator easily gripped between the thumb and forefinger. Letters were selected by placing a guide on the bottom of the indicator into the appropriate hole in the indicator plate, and pressing down. This forced the corresponding type on the rubber index down against the page and automatically moved the assembly along to the next space. Paper was controlled by a platen of small diameter and printing was blind—the indicator assembly could be raised, however, to make the typing visible.

It was well made, it was simple, it was small and light, it was elegant and it was relatively cheap. Over the years it proved immensely popular and sold in many, many thousands despite its numerous and obvious drawbacks, not the least of which was the fact that unless the typist was endowed with bionic vision or virtually straddled the machine, the letters on the index were all but invisible.

72. Hall Type Writer, with added index (Courtesy of Bernard Williams)

Many manufacturers in different parts of the world jumped on the Hall bandwagon with thinly disguised copies, but few of them achieved the same degree of success. Later, when rivals began crowding the index machines out of the market, Hall introduced a type sleeve version called Century, but without success. He also offered a curious modified version of the original (72), with a white enamelled plate and an extended indicator clumsily fitted to the machine to help prevent his clients going blind selecting the letters, but this hybrid was a further attempt to extend the basic design beyond its natural useful life. It was futile—the machine belonged strictly to the 1880s.

You could say the same of another quite wonderful little instrument called Columbia (9, 73, 137), the brainchild of a prolific typewriter inventor by the name of Charles Spiro. This New York watchmaker fitted a fairly substantial knob onto an otherwise delicate vertical type wheel assembly geared at 90° to an indicator on a circular index. Turning the knob selected the desired letter and pressing down on it displaced the inking pad and typed the letter onto a platen of small diameter beneath it. This was a simple and effective design, but Spiro's ingenuity did not stop there. By means of a pawl and plate fitted to the back of the type wheel, the process of lowering it displaced the platen the space required to accommodate the width of the letter being printed. For the first time in a mass produced machine, full differential spacing was being offered—Jones had also tried it, of course, but can hardly be said to have mass produced, back in 1852.

73. Columbia Type Writer No. 2 (Courtesy of Sotheby's, London)

74. Folding Hammond (Courtesy of J. G. Foster Collection)

So there were two very small portable machines—the Hall and the Columbia—in amongst the giants almost from the start. The Hammond was not all that large but it could not strictly speaking be called a portable. Many years later, Hammond had to introduce an aluminum model in order to make any sort of impression on the portable market. Soon after that, they went even further by introducing a folding model **(74)** with a hinged keyboard which could be raised to give the machine a more compact rectangular profile when not in use, thus effectively making a portable out of a larger machine. The Hall and the Columbia were small and portable from the start, however, and their elegant wooden cases added to their appeal.

The Columbia, darling of modern collectors the world over, is one of those typewriters which looks so quaint these days that we are tempted to wonder whether it was ever really taken seriously. This it certainly was; many thousands were manufactured and sold over the years, and quite a number have survived to grace contemporary collections. The really rare examples are the two early models, of which fewer were manufactured. The original upper case only machine was quickly replaced by one which typed upper and lower using the clumsy and improbable expedient of mount-

ing a second type wheel in tandem to the first, upper case on one wheel and lower case on the other, with lateral displacement for selection of case **(21, 137)**. Had Spiro persevered with this design, his invention would certainly have experienced a quick and painless demise; as it was, he had the foresight to drop the thing almost immediately and substitute the now more familiar Model no. 2 (which was actually the third model, of course, not the second) in which both upper and lower case were fitted to a single wheel **(9)**.

A contemporary and even smaller portable to challenge the Hall and the Columbia was introduced at the same time and if specifications alone had been permitted to determine results, then it ought to have won the battle for the portable market hands down. This was a tiny up stroke type bar machine with a three row keyboard dating from 1884 and manufactured in limited numbers under the names Automatic and Hamilton. The little type bars struck underneath the platen, in the manner of the Remington, so that at least its action would have been familiar to potential typewriter clients. It also offered the unusual and sophisticated feature, for a type bar device, of differential spacing. Yet none of this was enough to save it from a premature demise.

Meanwhile, on the other side of the Atlantic there were the first signs of activity as well. Malling Hansen (**frontispiece, 4, 43 et al.**) had already been in business well over a decade before the next European competitor put in an appearance in 1882 with a machine called Hammonia (**18, 172**). This was a remarkable and quite unorthodox linear index device which might easily be mistaken for a slicing machine, even at relatively close quarters. A substantial blade with a conspicuous handle on top and type underneath traversed the page, sliding back or forth for letter selection. Pressing down on the handle printed the desired character on the paper beneath it. Given the relative sophistication of its American and Continental contemporaries, this German entry into the typewriter market was never likely to have any commercial impact worthy of note, but just enough were made and sold for a few to have survived…to stimulate the flow of adrenaline in typewriter collectors the world over!

The other German entry into the market is even more elusive. It was a remarkable design from today's perspective but failed to make any impact whatever in its day—in fact, it was to be quite a long time before Europe began to present any sort of challenge to US domination in the typewriter field. The machine in question was invented by Ernst Wilhelm Brackelsberg in 1884 and called Westphalia. It was as unorthodox a design as that of its other German competitor and even less successful, for although a few appear to have been made, the device was never manufactured in any commercial quantities.

That, more or less, completes the picture as it appeared in the first few years of the 1880s. A number of typewriters was on the market on both sides of the Atlantic but none was enjoying much success as yet. Remington was only just beginning to assert itself, selling a few thousand machines a year, which must have seemed like an impossible dream to the likes of Hammond, Hall, Spiro, Crandall, and the others. Certainly more Remingtons were being sold at the time than all the other makes put together, but the fragmented market was already looking chaotic—there were type bar machines, swinging sector, index, type sleeve, and type wheel machines, not to mention type plungers. It is this vast diversity in typewriter developments and inventions, more than anything else, which makes the history of the typewriter the fascinating story that it is. What other field can boast as varied and versatile a past as the writing machine?

Before watching the momentum of mass production really gather speed, we should go back and pick up the careers of the original Sholes and Glidden group, who were by no means out of it at this stage. They may have lost control of their original invention, but as the 1880s developed, their prospects began to look more and more attractive.

75. Underwood (Courtesy of This Olde Office, Cathedral City, California)

The first one to bounce back was none other than George Washington Newton Yost. Having successfully brought Densmore and Remington together on that momentous occasion which launched the Sholes and Glidden at the beginning of 1873, he subsequently owned a share of the retail agency which had such a difficult time selling the product to a disinterested public some years later. In 1880, Yost formed the Caligraph Patent Company with Walter Barron, James Densmore's step-son, and began building a machine he had designed together with Franz Wagner, who some years later was to invent one of the most famous typewriters of all time: the Underwood (**75**).

The Caligraph (**65, 131**) was manufactured by the American Writing Machine Company and was launched in 1881 as an upper case only machine, a second model the following year offering both cases. A significant feature of the Caligraph was a relatively large flat cover between the keyboard and the typist which housed the ends of the long key levers which, it was claimed, helped to lighten the touch and also provided a convenient surface on which a tired typist could rest his or her hands. It was a blind machine, like the Remington, and the platen had to be raised to make typing visible. The other significant feature is that from the second model onwards the Caligraph became what is known as a 'full keyboard' machine, with a separate key for each upper and lower case character. There was no shift for change of case, the upper case black keys were on the two sides of the keyboard, with the white lower case keys in the middle. This pioneering full keyboard design was soon adopted by some other manufacturers but it proved an unnecessary complication and was ultimately abandoned.

The indomitable Yost seemed set to make the Caligraph his crowning achievement when suddenly he broke away from the American Writing Machine Company and set off on a new venture. The reasons for the change are uncertain—he had to pay royalties to Densmore and Barron for some of the features on which they held the patents and maybe he found this difficult to swallow. Together with some new associates (including one Jacob Felbel, who some years later was to contribute to the design of the great Smith Premier (**96, 244**), Yost launched a truly remarkable machine in 1887 to which he gave his name (**53, 60**).

The Yost, like the Caligraph, retained the idea of a separate key for each upper and lower case character, but unlike the latter's full keyboard, the Yost had a 'double keyboard' consisting of identical rows of keys for each case. The Caligraph's six rows of keys had now grown to eight rows on the Yost. But that was only part of it, and the lesser part, at that. The most unusual feature of the Yost was a truly astounding trajectory which the pivoted type bars followed on their way to the platen. At rest against a circular inking pad in the type basket, the bars hopped off the pad, over and up onto the platen in a movement which has come to be known as the 'grasshopper' action. The venture proved a success, and the machine which bore his name outlived him by many years after his death in 1895.

Yost's was not the only grasshopper in typewriter history. There is another which is perhaps even more remarkable because the hopping takes place not within the bowels of the machine where the action is partially obscured, but on top in full view. It is a sight which, once seen, is never forgotten—most collectors recall vividly the utter amazement they experienced when they first saw it. The machine is called Williams, after its inventor, and it put in its first appearance in 1891. Horizontally positioned type bars radiate outwards from the centrally located platen at the top of the device, with a curved (later straight) three row keyboard in front **(76, 273, 274)**. When a key is pressed, the corresponding bar quite literally hops up from its position of rest against an inking pad and over onto the paper. Of course, with the type bars both in front of and behind the platen, the obvious problem (once again) was what to do with the paper. Williams' solution was an arrangement whereby the page was first rolled into a frame inside the machine in front of the platen, trained up and over the platen during typing, and curled up into another frame at the rear as the typing proceeded. It sounds absolutely terrible, and it was, and it ought to have doomed the design from the start, but in fact the Williams proved a reasonably successful machine despite this handicap and production continued through a number of models for almost twenty years.

manufacturing one which his would rival: the Fitch **(78)**, that fine down stroke machine which looks like half a Williams with a fright! The Fitch was not a grasshopper machine, but was a rather strange down stroke with the type bars fanning out obliquely behind the platen in a raised structure which gave the machine a quite remarkable and unmistakable profile. The eventual Williams connection is further evidenced by their comparable arrangements for the sheet of paper, which coiled up in baskets on both instruments.

76.   Williams 3

There were a few other grasshoppers as well, but none was as successful as the Williams. John Nevil Maskelyne of London built an absolutely marvellous one, in several models but limited numbers, from 1890 onwards. Full differential spacing was one of its major features—to combine that with a grasshopper was quite an achievement in itself. Not content with that, he produced a quite stunning version in 1897 in which the grasshopper actually performed a complete somersault on its way to the platen. And on this side of the Atlantic, the Jackson Typewriter Company also marketed a grasshopper in 1898. The inventor of the Jackson **(77)** was Andrew J. Steiger, one of the group of men who collaborated in designing the Yost.

Here they are again: those recurring threads joining early machines and inventors to each other! Nor does it stop there. At the same time that Williams was perfecting his own machine, he was using the services afforded by a factory which was actually

78. Fitch (Courtesy of Christie's South Kensington, London)

In fact, coiling the paper in this improbable manner was a feature common to a number of like machines. They resorted to this system in their utter inability to dispose of the paper in any other way, as we saw on the Williams in particular, given the type bars on both sides of the platen. Hammond was the first to encounter this problem, as we have seen, and it was he who resorted to coiling up the unused page and allowing it to feed out unhindered. Machines like the Fitch which conformed to the down stroke

79. North's

from the rear design faced a different sort of dilemma. In theory the page would only have needed to be curled up at one end, except that the free end would have obscured the keyboard either on its way into the machine, or on the way out, or both. There were quite a number of these down stroke machines in which the type bars were positioned behind the platen and all of them faced the same problems with the paper. They are among the most attractive and desirable of type bar designs, to our late 20th century eyes, although whether they were considered as such by contemporary typists who had to fumble around with the rolling and unrolling of their sheets of paper is less certain. Given that none of them was ever what one might call a great commercial success, the typists may in fact have had the last word.

One such machine was designed by typewriter pioneer Byron Brooks, the Remington employee who helped remodel the Sholes and Glidden prototype in the early 1870s. He also held the patent on change of case so that Remington (and others) had to pay him royalties when they introduced their upper and lower case models later on. When he eventually went into production to market his own machine, (his 1885 patent beating Fitch by a year) his three row keyboard instrument with its semicircle of type bars positioned vertically behind the platen, although not solving the problem of the paper, set a style which the others more or less followed. Two Englishmen called Morgan Donne and George Cooper invented a comparable instrument in 1890 which appeared on the market two years later financed by Lord North and, predictably enough, named

North's in his honour **(79, 201)**. It was a fine and well made machine which deserved greater popularity than it enjoyed; even so, several thousand were manufactured over a number of years. It was in fact far more successful than Donne and Cooper's other effort, a down stroke machine named English with the type bars in front of the platen.

Interestingly enough, although the North's offers no choice but to roll the sheet of paper up like the other designs of its class, its inventors stuck with the idea on their English as well, even though there was absolutely no need in this instance as the type bars were in front of the platen. Some habits just seem impossible to break.

Down stroke machines with the type bars in front of the platen were in fact one of the most numerous categories and many of them, unlike the English, enjoyed great commercial success. Before looking at some of the most famous of them however, we should first consider one more down stroke machine with the type bars behind the platen which has not yet been mentioned. This remarkable contemporary of the others of its kind was called Waverley, patented in 1889 and manufactured briefly in London some years later **(80)**. What distinguished it from the others more than anything else was that it incorporated not only differential but also terminal spacing, making it very beautiful and desirable, despite the inevitable problems with the rolled paper entering and exiting the machine.

80. Waverley (Courtesy of Bernard Williams)

81. Hammond 12

All in all, the down stroke machines with the type bars behind the platen enjoyed only a limited success. However their siblings with type bars in front of the platen were a different story altogether and numerous makes and models were mass produced in their hundreds of thousands. Only one plausible reason can explain their vastly different fortunes: each concept had its own advantages and disadvantages, so in the end the bonus of visible text offered by typing from the rear was less commercially attractive than the ease of paper disposal offered by the down stroke from the front.

We have examined the paper problem in some detail. The page had to be coiled into a basket on all the down stroke designs with type bars behind the platen, and it would seem that no amount of ratiocination could overcome the handicap. Down stroke machines with the type bars in front of the platen provided no obstruction to the sheet of paper but the *quid pro quo* was that typewritten word was completely obscured unless the machine was on the floor or the typist was on her feet at the desk. Given the success of this design, the feature of visibility of type seems to have been strangely irrelevant, and the reason may well be that the popular contemporary up stroke competitors such as the Remington were also typing blind, and doing quite nicely on it. It simply did not seem to matter at first whether you could see what you were typing or not and, even though typewriters producing visible work were easily obtainable from an early date, their blind competitors had the market virtually to themselves until the hugely successful Underwood **(75)** almost single-handedly put the matter beyond dispute by becoming the prototype of the standard manual upright for more than the following half century.

Of the down stroke machines with the type bars in front of the platen, the English is pretty well unique in retaining the paper rolling feature. In any case its impact on the market was utterly insignificant since so few were produced. With clearly unintentional humour, the publicity for the machine promoted the feature of the rolled up page as highly desirable in preventing the inquisitive from surreptitiously reading the text over the typist's shoulder. Unfortunately, there were apparently not enough secret agents buying typewriters at the time to ensure its commercial success. In fact, only one manufacturer ever overcame the paper basket prejudice to a notable extent and that of course was the spectacularly successful Hammond, which we have already discussed, and this particular instrument did not use type bars. Its other virtues were so great anyway, that it alone was able to overcome residual resistance to rolling up the sheet of paper.

The popular down stroke machines with type bars in front of the platen, as well as their closely related oblique front stroke cousins, in which the type bars are not actually vertical but inclined away from the platen and towards the typist for greater visibility of typing, not only appeared very early on but also persisted well into the 20th century. The very first of these to appear was actually of the oblique variety, which should have ensured its success. In any event this machine, patented in 1883 and manufactured by the Horton Typewriter Company of Toronto **(82, 175)**, was a commercial failure and only a very few examples were made.

Another machine with vertical type bars in front of the platen was the Franklin **(83)**, invented in 1889 by that famous typewriter pioneer Wellington P. Kidder who was to enjoy such fame and fortune as the inventor of the thrust action machines to which he gave his Christian name, as well as the numerous makes—including the Noiseless **(84)** and Empire **(85, 157)**—which evolved from it. These were as spectacularly popular as Kidder's Franklin was relatively unsuccessful. Even though production continued for several decades the machine was never a best seller, hampered possibly by an ungainly curved, almost semicircular, keyboard.

82. Horton (Courtesy of Dan Post Archives)

83. Franklin (Courtesy of This Olde Office, Cathedral City, California)

84. Noiseless (Courtesy of This Olde Office, Cathedral City, California)

Numerous oblique front stroke machines, such as the Polygraph **(86)**, Norica, Triumph **(87)**, Bing, Celtic, a late model Salter **(88)** and even a folding Remington portable **(89)**, appeared on the scene over the years until well into the 1920s. Three of these machines, however, tower above the others in this category and deserve separate consideration.

The most important was the product of the great inventor Charles Spiro, the New York watchmaker who designed that beautiful little type wheel Columbia back in 1885. Four years later, he patented and began manufacturing a fine down stroke machine with type bars in front of the platen and with a double keyboard.

He called it Bar Lock **(90, 118, 136)** in honour of its semi circle of pins which were claimed to lock the type bars into perfect alignment and prevent them from jamming. It actually did neither, but this did not appear to matter very much to the public who purchased this excellent machine in many hundreds of thousands of units around the globe, before it became an oblique down stroke in 1911 in deference to popular demand, and ultimately a conventional front stroke from 1921 on **(118)**. Spiro meanwhile accumulated as many different patents for minor improvements in design as possibly anyone in typewriter history.

85. Empire (Courtesy of Bernard Williams)

86. Polygraph

87. Triumph (Courtesy of Milwaukee Public Museum)

88. Salter Visible (Courtesy of Bernard Williams)

89. Remington Portable (Courtesy of Bernard Williams)

George Salter, the famous British manufacturer of scales and weighing equipment, did not enjoy quite such an illustrious career. His down stroke device was based on an 1892 patent with the type bars located in a vertical semicircle like the Bar Lock but controlled by a small curved three row keyboard with double shift rather than the latter's large double keyboard. The device was not an immediate success by any means, probably because the manufacturers seemed to have an inordinate amount of trouble with their inking system, among other things. The actual patent protecting the in-

90. Bar-Lock (Courtesy of Bernard Williams)

vention specified an inking pad with the type bars resting against it and pirouetting in a most improbable manner as they approached the paper. This was an ungainly arrangement which was mercifully (for them…but not for us!) never manufactured although some machines, notably the Maskelyne and even the Sun **(8)** and Williams and Yost, were successful in combining type bars with unconventional inking systems based on pads and rollers.

By the time the Salter reached the market, however, this patented pad arrangement had been dropped in favour of an inking roller against which the type bars rubbed on their way to the paper. It was clearly doomed, too, and was soon replaced by a ribbon with subsequent minor modifications to width, advance, and so on. Success at last **(22)**. Numerous model changes ensued, the curved keyboard giving way to a straight three row arrangement in 1900 **(91)** and then to a conventional four row version in 1913, when the vertical type bars were tilted back towards the typist **(92)** much as they had been on the Bar Lock two years earlier.

91. Salter 6 (Courtesy of Christie's South Kensington, London)

92. Salter Visible (Courtesy of Christie's South Kensington, London)

The other famous manufacturer of a comparable instrument was the Imperial Typewriter Company in Leicester. Their machine had a chequered career, beginning life in 1903 as the type sleeve Moya **(93, 195, 196)**, named after its inventor and manufacturer Hidalgo Moya. Five years and several models later, with commercial failure staring him in the face, the inventor and his colleagues changed course and placed on the market a down stroke machine with a curved three row keyboard. This was named the Imperial **(94)** and it proved as resounding a success as its predecessors had proved to be qualified failures, despite the Imperial's obvious competition from conventional front stroke machines. Down stroke Imperials were made in England and on the Continent until the early 1930s, even vying with the same company's later conventional four row front stroke models. Introduced in 1927, it was with these front stroke machines that the manufacturer ultimately carved out his main niche in the market for several decades to come.

93. Moya Visible No. 2 (Courtesy of Christie's South Kensington, London)

94. Imperial

Well then, having now briefly sketched in the association between many of these early ideas and their later developments, let us return once again to Sholes, Glidden, and Remington and the world-wide industry they were so instrumental in promoting, before we conclude by gathering up the loose ends.

The Remington organization, as we have seen, proved to be a veritable spawning ground of early typewriter talent. Some of its products, like Clough and Jenne and Washburn, made their individual contributions and then passed into relative obscurity. Others maintained a much higher profile—Yost, Brooks, and Hammond, for instance.

And Henry Benedict.

Benedict started off as a Remington employee and eventually became a director of that company and one of the men responsible for the infant Type Writer. When Remington considered divesting themselves of their largely unsuccessful typewriter division, Benedict was one of the men ready to step in. He was joined by a keen and devoted typewriter advocate called Wyckoff who was not only an early user of the machine itself but who had collared the retail agency for Central New York State back in the days when the going was tough and sales were dismal, forcing every member of his staff to learn how to type or quit. They typed! Benedict and Wyckoff were joined by a third ex-Remington employee called Clarence Seamans, a dynamic individual active in the retail side of the business.

The three men formed the partnership of Wyckoff, Seamans and Benedict in 1882 and the success of the typewriter dates from this historic event. Sales were slow at first, but by 1885 they had expanded dramatically and ultimate success was ensured.

Remington itself, however, was sinking deeper into financial crisis which threatened the entire enterprise, and Philo Remington was ready to sell the typewriter division. He was desperate for money to satisfy the more persistent of his creditors and Benedict was ready to put cash up front. The result was inevitable and the retail partnership of Wyckoff, Seamans and Benedict bought the entire typewriter division, together with all the patent rights, franchises, and manufacturing facilities. When Remington and Sons collapsed shortly afterwards, the typewriter division was thus saved and continued to use the name even though the Remington family had nothing further to do with the company. The man in charge of manufacture was W. K. Jenne whose involvement with the machine dated back to the very earliest days, while sales and distribution continued in the capable hands of Wyckoff, Seamans, and Benedict.

They were destined to go from strength to strength. The original Sholes and Glidden upper case machine was slow and clumsy and its limited sales reflected not only public apathy but also the indifferent performance of the machine itself. With the advent of the Remington Model 2 in 1878 however, the picture began to change. This model proved to be one of the truly great machines of typewriter history. Despite the fact that it followed the Sholes and Glidden's blind format, its familiar upright profile and its even more familiar keyboard imposed upon the world a design which was to dominate the market for decades to come.

Of the other pioneers who began their careers in association with Remington, further mention must be made of the Densmore brothers. James, Amos, and Emmett Densmore were all partners to different degrees in the original Sholes and Glidden enterprise, before divesting themselves of their shares through complex and ill-tempered dealings. James lived out his last cantankerous years collecting royalties and fighting legal battles against all the world. Amos and Emmett Densmore, however, joined forces with James' step-son, Walter Barron, and also with Franz Wagner who had earlier designed Yost's Caligraph (65) and was later to hit the jackpot with the Underwood (75). The product of their association was a

95.   Densmore 5 (Courtesy of Bernard Williams)

well conceived up stroke machine with four row keyboard, which they rightly called Densmore and which they marketed with a modicum of success (95) until the Union Typewriter Company acquired it in 1893 and eventually discontinued it in the early years of this century.

Another machine to be taken over by the notorious typewriter trust called Union Typewriter Company was the famous Smith Premier (96, 244). This somewhat cumbersome double keyboard design enjoyed huge success in its day, challenging the conventional keyboard machines by dispensing with the shift key and allocating a separate key to each upper and lower case letter. The result looked heavy and clumsy but the quality of the work it produced amply compensated for this failing, despite the fact that it was an up stroke type bar machine, typing blind like the Remington. It had some other novel features in its favour though: uniquely, a circular brush inside the type basket could be turned by means of a handle to clean the type face in situ, and its ribbon, like that of the earlier Brooks, zigzagged so that its entire width could be used, which was no small advantage in its day. And since we have already noted how incestuous early typewriter history was, it is worth mentioning that one of the men involved in the design of the Smith Premier was the aforementioned Jacob Felbel who a few years earlier had been largely responsible for the design of the grasshopper Yost.

While all these type bar designs were carving out their share of typewriter history, some alternative designs were making just as big an impact upon the market. The Hammond (64, 71 et al.), as we have already noted, was well on the way to establishing itself as one of the great machines of all time despite features which might have doomed a lesser design. Its unconventional treatment of placing the sheet of paper between the type sector in front of the page and a hammer behind it produced printing by striking the paper against the type sector, resulting in regularity of image unmatched on any other manual typewriter ever made. Between its original format and its later reincarnation as the great VariTyper (171) which was destined to play an important role in the printing industry, many hundreds of thousands of these machines were manufactured.

Other alternatives to the type bar which eventually imposed themselves upon the market in a big way were the type sleeve and type wheel. We have already examined the type sleeve as used by

96. Smith Premier (Courtesy of This Olde Office, Cathedral City, California)

Lucien Stephen Crandall on the machine to which he gave his name. The very limited success which this design enjoyed might well be attributed to the type sleeve as a concept, were it not for the fact that an even more unlikely invention using the same typing principle proved such a staggering commercial success. The machine in question was the Mignon **(47, 189, 190)**, and a less likely candidate for the typewriter best seller list would be hard to imagine. Manufactured in Germany from the early years of this century, it was in direct competition with the Remingtons and the Underwoods and other visible front stroke uprights right from the start so that its enormous success was nothing less than miraculous.

Not only did the Mignon use a type sleeve, which by its very design was less efficient than a type wheel, but it even did so without a keyboard! Instead it used an indicator suspended above a rectangular index so that the typist selected the desired letter by placing the indicator over it with the left hand and pressing a typing key with the right. The entire operation therefore required only the thumb and forefinger of the left and the forefinger of the right hand. A small rack attached to the indicator turned the type sleeve pinion to the desired letter, and pressure on the key brought the type sleeve down onto the sheet of paper. It was an improbable design for such a popular machine, and these days when a member of the public who has not been blessed by initiation into the

97. Royal

arcane and esoteric mysteries of typewriter history first encounters a Mignon, more likely than not he will be convinced that he is in the presence of the first typewriter ever made.

While the Mignon was selling in such numbers all over the globe, a more conventional machine was equally successful with a design destined to make a big come-back in the second half of this century. The Blickensderfer **(49, 63, et al.)** used a three row keyboard to control a small type wheel. Beautifully designed and made, despite the handicap of inking by roller, it was even more commercially successful than the Mignon. Both machines had the immense advantage over their type bar competitors of instant and easy interchangeability of type—literally hundreds of different type sleeves and type wheels for these machines were readily available and covered every conceivable alphabet and specialized technical and scientific requirement. In this respect the Mignon bettered the Blickensderfer (or the Hammond for that matter, which also offered as many different shuttles), because every Mignon type sleeve had a matching index which could be slipped into position for immediate selection of the desired character on its type sleeve. Nevertheless the Blickensderfer, which of course was by no means the only keyboard machine to use a type wheel, was vindicated some

decades after its demise by the introduction of a whole new generation of the modern electric golf ball instruments so popular in our own times.

While talking about best selling machines, one other particular type bar design deserves separate mention. In 1888 an American called Bernard Granville patented a design, which achieved limited manufacture some years later, in which type bars did not strike the paper as on conventional machines but rather were thrust against it. He called his invention the Rapid, although the only notable speed it achieved was that of its early demise. However, it launched a design which was to achieve enormous success on both sides of the Atlantic for many decades.

Four years after Granville's patent, another American, Wellington Parker Kidder, invented a thrust action machine with a three row keyboard in which the type bars were formed in an arc opposite the printing point and were thrust in a straight horizontal line against the platen when the corresponding keys were pressed. The machine was a phenomenal success almost overnight. Under its inventor's Christian name of Wellington, but even more under the name Empire, it sold in large numbers all over the world well into the 20th century **(85, 157)**. It was manufactured under licence in Germany as the even more successful Adler, with huge production continuing in that country through successive models long after it had ceased elsewhere in the world.

This was without doubt one of the all-time great designs of typewriter history and it was almost inevitable that anything Kidder attempted afterwards would be an anti-climax, and so it was with his Franklin **(83)** which proved more of a historical curiosity than a commercial success. A radical departure from the earlier thrust action effort, the Franklin was a down stroke machine with a curved keyboard of at first two, but later three, rows. Limited manufacture continued on a relatively small scale until the early years of this century.

But the thrust idea was clearly more commercial, and Kidder proceeded to exploit its relatively quiet action (since the keys actually pressed against the paper rather than slammed into it) in an ingenious application of the principle in an instrument appropriately called Noiseless **(84)**. A certain measure of commercial success crowned his efforts until Remington bought the company in 1924 and marketed the machines as the Remington Noiseless **(222, 223)**. Meanwhile, an American by the name of Ford had marketed a comparable thrust action instrument under his own name in the United States as well as in France and Germany (where it was named Hurtu and Knoch respectively), but without in any way threatening the domination of Kidder designed machines.

One failing of all these thrust action instruments was their allegedly poor performance in cutting stencils, which were becoming more and more popular in offices around the world for the printing of leaflets and circulars. The problem was inherent in the thrust design—even with overthrow weights, a thrust was not as good as a blow when it came to cutting stencils.

Mimeograph inventor Thomas Edison appreciated early on that to cut a crisp, clean stencil what you really needed was a hammer. Edison was already familiar with the use of a hammer, having fitted such a device to his printing telegraph. His typewriter **(17)**, patented in 1895, made good use of it once again. Type plungers were fitted vertically around the perimeter of a horizontal wheel in such a manner that pressing a key on the base of the machine caused a hammer to strike the selected plunger up against the platen and knock it deftly back down again as the hammer fell. It was not the most felicitous of Edison's many inventions…but it certainly cut clean stencils!

It was about this time that a design which was to change the course of our history made its dramatic re-appearance. Franz Xaver Wagner, a German born technician who emigrated to the United States at an early age, enjoyed a pedigree as impeccable as any man in the business. He had already achieved a modicum of success as an inventor even before he turned his attention to the development of the infant Type Writer in the Remington factory **(rear cover, 1, 12, etc.)**. Those who met him were clearly impressed by his talents. Yost used him to assist in the design of both the Caligraph **(65)** and the Yost **(60)** typewriters and the Densmores used his help with the machine to which they gave their name **(95)**.

Wagner soon realised, however, that the future lay with a machine which offered visible typing. The outcome was a front stroke upright which became the prototype of all conventional typewriters for the next half century. An abortive attempt at manufacture by a company which he himself formed to promote the invention was abandoned when a ribbon manufacturer called John Underwood bought him out…and the rest is Underwood history.

So successful was the Underwood design **(75)** and so totally did it impose itself on the world market that it was not long before every major manufacturer on both sides of the Atlantic produced machines in its image. Despite this competition, sales of the Underwood Model 5, which represented the pinnacle of the company's success, were estimated to account for about 50% of the total typewriter market in the 1920s, a quite staggering success which no other machine has ever matched. And yet it was this very pre-eminence that contributed to the company's eventual decline, for while it was content to rest on its laurels, its competitors began developing improved machines and designs which eventually displaced it from its pinnacle.

There are other names as well, which do not merit individual mention but which some of us of a certain age will remember with some nostalgia. These were the uprights ubiquitous between the Wars, names such as Royal **(97)**, Woodstock **(98)**, Smith, Stearns **(99)**, Torpedo, Olivetti, Mercedes, Halda, Kappel, Continental **(100)**, Demountable **(101)**…and there are many, many more.

98. Woodstock (Courtesy of This Olde Office, Cathedral City, California)

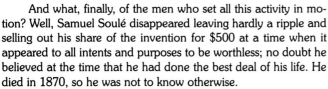

99. Stearns (Courtesy of This Olde Office, Cathdral City, California)

100. Continental (Courtesy of This Olde Office, Cathedral City, California)

And what, finally, of the men who set all this activity in motion? Well, Samuel Soulé disappeared leaving hardly a ripple and selling out his share of the invention for $500 at a time when it appeared to all intents and purposes to be worthless; no doubt he believed at the time that he had done the best deal of his life. He died in 1870, so he was not to know otherwise.

Carlos Glidden lived on until 1877, so he was around just long enough to realise that their early efforts had not been in vain. He had also sold his share early on for a mere $250, then had second thoughts and claimed back a tenth interest which Sholes ceded to him. This is perhaps not quite the stroke of magnanimity it appears, because Sholes himself was convinced at the time that the whole enterprise was worthless and of course (bethought Sholes to himself) ten per cent of nothing is, after all…nothing.

As for Sholes himself, it was only thanks to the self-interest of James Densmore that he finished up making anything at all from the invention. It was Densmore who prevented Sholes from selling out, time and time again, not out of love but out of necessity because he needed him. Even long after the Type Writer had outgrown them both, Sholes himself and several of his sons were still busy inventing other machines such as the ugly but highly original Sholes Visible **(102)**, but with mixed success. None of them was to match the first effort, however, and despite the gloom and doom with which he had viewed the enterprise in those early difficult days, Sholes was able to write shortly before his death in 1890:

'Whatever I may have felt in the early days of the value of the typewriter, it is obviously a blessing to mankind.'

Chapter Five—
Directory Of Inventions
Never Manufactured

ADDEY

An 1889 machine which sported a very futuristic feature: '…letters placed around a small ball or sphere, 2 inches in diameter, which is caused to revolve. The letters themselves are ranged in six lines converging at the poles…'[62] Addey's Typograph was not the only golf ball in its day by any means, nor even the first (see also Richardson etc.), but a type wheel of spherical design was a more sophisticated concept than the more common cylinder. The platen of the Typograph struck against the type wheel. It had thirty-six keys printing three characters each with the help of two shift keys which, with a spacer, were located in the centre of the keyboard.

ADERS

A pneumatic typewriter using compressed air or gas was patented in Germany in 1907 by Franz Aders but the project (thankfully) never materialized.

ALBERTSON

Ernest Albertson designed a folding front stroke machine with a four row keyboard, patented in 1913. Due to be manufactured by the Albertson Typewriter Company, it was never marketed.

ALISSOFF

A Russian inventor by this name designed and built a type wheel machine, patented in France in 1872 and in England two years later. An example was exhibited at the Philadelphia Exhibition in 1876 where it competed with the recently introduced Sholes and Glidden model made by Remington. It won this particular competition, receiving eloquent commendation in the official reports for the beauty of its construction and design while the American machine, which was crude by comparison, was not even mentioned. History can make judges look foolish and one can only assume that they were swayed by the superior quality of the construction since the details of the design were somewhat clumsy and anachronistic. The letters of the alphabet, upper and lower case, were displayed on a circular index and selected by means of a handle attached to a type wheel. Several alphabets, each on its own type wheel, could be mounted concentrically and the appropriate one selected by turning a handle attached to a worm at the bottom of the machine. Paper was fed around a cylindrical platen and lowered onto the type by means of a handle, inking by roller.

ALLEN R T P

The initials of the name are supplied to avoid confusion with a later machine. Two US patents were assigned to an inventor of this name but neither was ever manufactured. The first, in 1875, was for a plunger machine with a circular keyboard, the plungers arranged in concentric circles and bearing more than a passing resemblance to the earlier Hansen Writing Ball.

A second patent was granted the following year for a different design altogether but which fared no better than the first. It now sported a conventional five row keyboard, type bars, and up stroke action.

ALPHABETAIRE TYPOGRAPHIQUE

A Frenchman by the name of Cabanel was granted a patent in 1875 for a type wheel machine with a letter index which was never manufactured.

AMBLER

As late as 1884 anachronistic patents were still being granted, this one to J. A. Ambler of Massachusetts for a design consisting of three concentric circles of type plungers striking down on a platen, with ribbon inking.

AMBROSETTI

Valerio Ambrosetti in 1883 was responsible for an improved version of Mazzei's Stenotiposillabica. It was allegedly capable of up to 220 wpm but was never manufactured.

AMERICAN DOUBLE WHEEL

Another of the remarkable designs in which Lee Burridge had a hand was a patent he shared with N. R. Marshman and assigned to the American Typewriter Company. Dated 1898, this hybrid sported two swinging type sectors one above the other and each controlled by its own arm. Letters were selected by placing the ring at the end of this arm over the desired key, the operator's left and right hand each operating its own lever and type sector. Printing was performed by pressing down the ring and key. It had pad inking. This intriguing design was clearly doomed.

AMERICAN POCKET

A small portable typewriter with a three row keyboard by this name in Milwaukee is reported to have been invented in 1926 by A. M. Leggatt of New York.[42] However a similar instrument described as a Leggatt by Typewriter Topics was produced three years earlier by Rochester Industries.[92]

ANDERSON H & F

Type plungers on a swinging sector attached to an indicator arm were the features of a patent granted to H. and F. Anderson of New York in 1878. It was an up stroke machine with the plungers swinging beneath the platen.

ANDERSON, J. TATE

A revolving type wheel controlled by a pin barrel was patented as late as 1877 by J. Tate Anderson of Philadelphia but was not manufactured.

ARNOLD

A patent for a swinging sector instrument with a curved letter index was granted to Benjamin Arnold of Rhode Island in 1876. Only a patent model has survived.[42]

AUTOMATIC

A grandiose scheme which never got off the ground. Proposed in 1929, the Automatic Word Writing Machine was designed to type not only in the conventional manner but, using 164 word keys, to print whole words at a single stroke.

AXIAL

The Axial Typewriter Company of Maine never apparently manufactured the two designs patented by J. H. Waite of Massachusetts in 1887 and assigned to it in the same year. One

101. Demountable (Courtesy of This Olde Office, Cathedral City, California)

was a circular index machine, the other employed a circle of type bars.

AZEVEDO

A Brazilian priest by the name of Padre Francisco João de Azevedo exhibited a typewriter of his invention at a National Exposition in Recife in 1861 and in Rio de Janeiro later the same year. Separate versions of the same instrument are known to have been built as a typewriter and as a shorthand machine, the principles of both being the same.

Azevedo's machines used the unusual principal of clamping the paper between two keys to produce an embossed image in a manner more commonly associated with Beach's 1856 invention. The keyboard consisted of twenty-four keys, of which only sixteen were used on the stenographic model. Paper tape more or less three fingers wide received the impression—combinations of letters (CP = G, DB = H, and so on) made it possible to complete the alphabet. The case was elegantly made of rosewood—everything in the machine with the exception of the type face and springs was also of wood. Azevedo lived until 1880—long enough to see the successful commercial manufacture of the typewriter. **(23)**

BADEAU

Isaac Badeau of New York was granted a US patent for a musical typewriter in 1903 but neither this design nor later ones patented jointly with his brother Louis were ever manufactured.

BAILEY

A shorthand machine subject of an 1896 US patent used thirty-four keys to perforate a shorthand code on paper tape. Several keys could be pressed simultaneously.

BAILLET de SONDALO and CORÉ

Two Frenchman patented a type wheel machine in 1841 which promised 'revolution radicale dans l'art typographique' that would make the pen obsolete and correspondence simple, even for those who did not know how to write. In the event, their principal contribution to typewriter history may well be their two incidental suggestions: that their invention could be

adapted for use in typing whole syllables at a time, and that by transposing type a message could be printed automatically in code. Both these concepts were destined to play a considerable role in the future history of the typewriter.

Syllable typewriters—those in which a single key printed the two or three letters of a syllable in one movement—were produced in considerable numbers late in the 19th century and some of them proved very good machines indeed. The fact that they all finally failed was because they did not offer the increased typing speed for which they were designed. They were bigger and more complicated (one even sported 132 keys, set in four keyboards) but without any discernible advantage. The printing cipher machines had a different history altogether and became progressively more important as the codes they were capable of printing became ever more complex. There were literally dozens of designs developed in the course of the century of which Baillet de Sondalo and Coré's was the first. Unfortunately for their invention and for those which followed, the simple substitution codes which they were capable of printing could be broken even by a child. This does not alter the fact that they were the first to suggest that by altering the relationship between type wheel and index, the machine would automatically print a coded message.

On this invention, the paper was positioned around a cylinder covered in soft leather (a very early application of the cylindrical platen, invented a few years earlier by their countryman Bidet). Letters were selected by an indicator on the right and a pedal brought type wheel and paper together. Line and letter spacing were performed manually by a process of considerable complexity. A second pedal could be adapted, if required, to change the relationship between index and type wheel for encoding. By progressively enlarging the design, the type wheel could allegedly carry up to 500 syllables and as many as six letters could be printed simultaneously.

BAIN

If we limit ourselves to stating merely that Alexander Bain's principal contribution to our history was his invention of the typewriter ribbon in 1841, we would be doing him a grave injustice. Whole books could be (and have been) written about the man and his many inventions and viewed from this perspective, the typewriter ribbon does not rate very highly on the scale of importance when considered against the fact that it was he who invented (*inter alia*) the electric clock and who made many valuable contributions to the history of telegraphy. His electro-chemical telegraph proposed the discoloration of chemically treated paper tape by longer and shorter electrical impulses according to the code in illustration. **(52)** Bain's patents of 1841 include designs using both type bars and type wheels, both of which had already been invented. Not so the ribbon, however, which his patent no. 9204 of 1841 spells out loud and clear: 'a roll of half inch wide ribbon, which is rubbed over the surface with a composition of two parts oil, four parts lamp black and one part spirits of turpentine, which will enable the pressure of any form thereon, as the type, to communicate the same to the paper.' Two years later he patented an endless ribbon, while in 1845 he completed the process by having the ribbon wound from one spool onto another, as typing progressed.

BALDRIDGE

W. D. Baldridge was granted a US patent in 1886 for an electrically assisted type bar machine with a four row keyboard. This did not reach manufacture.

BARLOW

A British patent for a primitive linear index device was granted to Alfred Barrow in 1874.

BARRATT

The Barratt Typewriter Company of London was the promoter of this 1915 product, allegedly with British Government backing, to manufacture the pre-war German Stoewer in England.

BARRETT

One of the inventors of the Fox typewriter, Glen J. Barrett produced the prototype of a visible machine which he had planned to market in 1904 under his own name but which was eventually marketed as the Fox.

BATH

The Londoner John Bath was granted a patent in 1869 for a type bar machine with a keyboard of forty-eight keys typing on a flat paper table.

BAUDOUIN & BOMBART

These two Frenchmen invented a shorthand machine with twenty keys in 1876. Printing was on paper tape by means of a ribbon, with the characters on flexible metal strips arranged in a swinging sector.

BEACH

Alfred Ely Beach made several machines of which the first appeared in 1847. He was later declared to be the inventor of the original typewriter in the pages of the *Scientific American*, a claim no doubt attributable to the fact that he was its editor and part owner. His first invention was a type bar device with a keyboard and flat paper using carbon paper for the impression but it 'did not satisfy the inventor's eye.' His second machine, patented in 1856, was a large device the size of a desk top with a three row keyboard and two type bars for each letter which clamped over the paper simultaneously to create an embossed impression—the only up and down stroke machine ever designed. A thin paper tape was used and no provision was made for typing on sheets, although impression through a strip of carbon paper was also suggested. In some respects, however, the machine was quite sophisticated, for it used a spring-driven escapement and a universal bar.

BECKENRIDGE

An 1880 US patent was granted to a J. W. Beckenridge for a machine with four keys per hand printing the alphabet by simultaneous depression of one or more keys. It could also be used as an attachment to existing typewriters.

BEETHOVEN

Several attempts at producing a musical typewriter by this name were made by Louis Aillaud of France from 1928 on. The first models were based on the Smith Premier no 10, with a short piano keyboard of eighteen keys attached to the front.

BELLOWS

An 1878 US patent was granted to Benjamin Bellows for a machine with a three row keyboard, printing by means of a hammer striking type against the platen.

BENNINGTON

An attempt at producing a syllable machine was made by W. H. Bennington of Kansas City in 1903. The down stroke design had a five row keyboard of sixty-eight keys, some of which typed short, commonly used words. Nothing came of the scheme, but many years later a similar design was marketed for a limited time as Xcel.

BERGER

A blind man called Octave Berger invented a machine in 1919 with six keys embossing a sheet of paper around a cylinder.

BERRY

A musical typewriter of considerable complexity was patented in 1836 by an Englishman named Miles Berry. Designed to be attached to the underside of a piano keyboard with a separate linkage for each note, the instrument used carbon paper to print a musical code of longer or shorter lines on paper advanced by clockwork. The length of the lines was determined by the length of time the piano keys remained depressed. Despite the passage of time, it failed to resolve technical

102. Sholes Visible

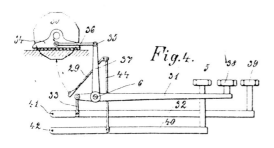

103. Brade (French patent)

BESERMENJE

Michael Besermenje spent seven years developmental work on a Yugoslav musical typewriter which he invented in 1925, but it was never manufactured.

BEYERLEN

A writing machine using six keys to emboss braille cells was invented in 1882 by a German called Angelo Beyerlen who was to enjoy a long and lucrative career as the Remington (and later Yost) representative for Germany and Austria.

BIANCHI

Another attempt at a musical typewriter was a Gravicembalo designed by an Italian inventor of this name in approximately 1847. The exact details and dates of the machine are not known, other than it sported a piano keyboard, type bars, and a cylindrical platen printing along the platen and rotating for line spacing as on modern machines. (Many early instru-

ments using cylindrical platens printed around the platen rather than along its length.)

BIDET

The French clock maker Gustave Jean Emile Bidet patented an interesting Compositeur Typographique in 1837, only to be stripped of his patent by royal decree some three years later.[27] His place in history is assured nonetheless, by virtue of the fact that his design was the first to incorporate a cylindrical platen.

A type wheel was mounted on the frame above the platen which was raised to bring it into contact with the type. Printing was around the cylinder which was turned by another small handle, with provision for differential spacing and inking by rollers on either side of the type wheel. If Bidet is not credited with the first typewriter to incorporate differential spacing this is due to the fact that it was not mechanically performed. Worse designs than his were still gracing the shelves of patent offices sixty years later, and type wheels with inking rollers were being manufactured all over the world until well into this century.

BLAKE

C. Waldon Blake and his Blake Typewriter Company of New-ark, New Jersey, planned in 1905 to resurrect the recently defunct Manhattan, which was itself a re-animation of the old Remington Model 2. The project was in the hands of the receiver within four years.

BLIND

Webster's 1898 Catalogue mentions a US machine by this name but details are unavailable.

BLITZ (1)

One of the inventors of the Kneist typewriter, Otto Ferdinand Mayer, was reported to have designed a type wheel device by this name, but it was not manufactured.

BLITZ (2)

An up stroke type bar syllable machine patented in Germany in 1896 by Richard Toepper, due to be manufactured under this name but the project did not materialize. The keyboard was to be increased by an additional three rows to handle the most frequent syllables and short words.

BOCQUET

An electrically operated device consisting of a standard typewriter wired to one for the blind was unsuccessfully attempted in 1923 by a blind Frenchman called Bocquet.

BOOK ELECTRIC

S. V. Clevenger invented several machines, one of which bore his name, but none of his inventions was commercially successful.

BOSCH

Robert Bosch, whose name was to become a household word in electrical equipment of all kinds, designed a typewriter for the blind in 1893. Similar to the Hall Type Writer, it selected Braille cells by means of an indicator.

BRACKELSBERG

Ernst Wilhelm Brackelsberg, who patented a linear index machine called Westphalia in 1884, made a further contribution thirteen years later with a syllable typewriter capable of printing up to ten letters simultaneously. The idea was to assemble keyboards side by side, each operating its own type sector so that simultaneous depression typed syllables or words in a single motion. The device was not manufactured.

BRADE

1902 French patent granted to Julius Brade for a stenographic system based upon letters placed above, below, or on a line to indicate vowels and for the modifications necessary to convert existing machines—the patent lists such machines as

Hammond, Pittsburg, Yost, and *Jewett* as suitable for conversion—by using additional keys and a rocker arm in order to position the character, as required, for selecting the desired vowel. The top row (Fig. 1) of the patent illustration **(103)** reads: 1) *nid*, 2) *gout*, 3) *dame*, 4) *chic*, 5) *bruit*, 6) *marche*, 7) *constriction*.

BRADFORD

The Electric Power Typewriter Co. of Bradford, Canada, announced the production in 1906 of an electric syllable machine with 300 keys but the project never materialized. The battery operated device was to cost $600, a considerable sum at the time, but a threefold increase in typing speed was assured and the machine was claimed to be fool-proof.

BRADY & WARNER

A United States patent granted in 1878 for a circular index machine was not manufactured.

BRAMAH

Joseph Bramah was a most prolific inventor, best remembered for the padlock which bears his name. Inspired no doubt by the versatility of his concentric wheels, he designed a primitive printing device using the same principle, for which he was granted a patent in 1806. Labrunie de Nerval developed the same idea some decades later, but both devices are more related to printing than to typing.

BROWN

C. T. Brown of Chicago patented an interesting grasshopper design in 1879 with the keys positioned above the type bars, but it did not attract a manufacturer.

BROWN-WOOD

An index design with type on the end of steel teeth above a cylindrical platen was said to have been invented by A. Brown-Wood of Michigan in the 1890s but the report has defied all efforts at confirmation.

BRYOIS

Monsieur Henri-Léon Bryois first patented his Sténographe Imprimeur in 1864. The complex design permitted the use of as many keys as were required—rows and rows of them—printing on paper tape advanced by clockwork. Two years later he patented a second design with 'divers perfectionnements' although these were never identified.[26] With a keyboard sporting a grid of 360 type plungers, the machine was by now so complex that no one other than the inventor was prepared to tackle it. It is rumoured that he ultimately redesigned the machine limiting the number of keys to twenty but there is no evidence that it ever left the drawing board.

BURBRA

A Mexican by the name of Juan Gualberto Holguin is reported to have invented a pneumatic machine by this name in 1914.

BURG

A type wheel machine capable of the simultaneous printing of a cipher text and a plain text was invented in 1906 by Hubert Burg of Germany.

BURLINGAME

An automatic printing telegraphic typewriter was invented by this American and patented in Britain by the Burlingame Typewriter Co. of San Francisco in 1908. Production was to start the same year, but the project failed to materialize.

BURNHAM

A prototype of a $300 musical typewriter invented by this resident of New Jersey was exhibited at the 1904 St. Louis Exhibition but the machine was not manufactured.

BURRIDGE

The fertile inventive genius of Lee S. Burridge of New York gave birth to some remarkable designs, among which is this

1897 US patent for a rocking radial type plunger machine in which each plunger had three keys at the upper end and nine characters at the lower. It never left the drawing board.

BURRY PAGE PRINTING TELEGRAPH

This was an attempt by New Yorker John Burry to design a telegraphic machine that would print messages to be torn as pages from an endless roll rather than on paper tape which needed to be glued to a page. With roller inking and printing performed by a travelling type wheel, the performance was equivalent to that obtained on any typewritten page.

BURT (1)

The first US patent for a writing machine was granted on 23rd July 1829 to a resident of Detroit called William Austin Burt. He called his invention a Typographer and although it was tragically destroyed by fire in 1836, a letter he typed on the machine has survived. For a long time it was considered the earliest known example of typing, although that honour is now conferred on the Turri correspondence.

Burt designed two machines which differed considerably in appearance but were similar in design. Both employed the swinging type sector principle for the first time, and Burt may justifiably claim to be the inventor of a concept which was to feature prominently in many later 19th century instruments. Essentially, the machines consisted of an arm pivoted at one end and bearing an arc of type, with a pointer and letter scale, at the other end. Swinging the arm to the correct letter on the scale brought the corresponding type over the printing point with pads on either side, over which the type sector brushed, providing the inking. Burt chalked up another first by offering upper as well as lower case on his sector. A roll of paper was specified, the end of which was held in a flat frame, and the typewritten sheets were torn off the roll at lengths indicated by a clock gauge on the front of the instrument. The Typographer produced quite acceptable work but it was slow, and despite a second improved model, its performance remained lethargic and failed to attract business interests. **(55)**

BURT (2)

An 1885 patent for a circular index machine with type on segmental teeth was granted to S. Burt of Chicago. Numerous such designs were common to different parts of the world at the time.

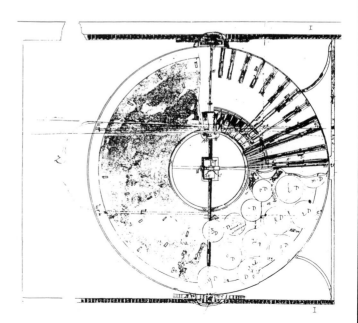

104. Codvelle (French patent)

II. *A Letter from Mr.* John Freke *F. R. S. Surgeon to St.* Bartholomew's *Hospital, to the* President *of the* Royal Society, *inclosing a Paper of the late Rev. Mr.* Creed, *concerning a* Machine *to write down Extempore Voluntaries, or other Pieces of Music.*

S I R,

Read March 12. 1746-7. I THINK the inclosed Paper is the Effect of great Ingenuity and much Thought; and as the Subject-Matter of it may tend to give great Improvement and Pleasure to many, not only in our own Country, but every-where, I hope my presenting it may not be thought improper that it may thereby be printed and published to the World.

It was invented and written by Mr. *Creed*, a Clergyman, who was esteemed, by those who knew him, to be a Man well acquainted with all kinds of mathematical Knowlege. It was sent me by a Gentleman of very distinguished Merit and Worth; if therefore from hence this Paper shall be thought proper to be published in the *Philosophical Transactions*. It will prevent its being lost to Mankind. I am,

S I R,

Your very humble

and obliged Servant,

John Freke.

M m m 2   A

A Demonstration of the Possibility of making a Machine *that shall write* Extempore Voluntaries, *or other Pieces of Music, as fast as any Master shall be able to play them upon an Organ, Harpsichord, &c. and that in a Character more natural and intelligible, and more expressive of all the Varieties those Instruments are capable of exhibiting, than the Character now in Use.*

Maxim I.

ALL the Varieties those Instruments afford fall under these three Heads: *First*, The various Durations of Sounds, commonly called *Minims, Crotchets,* &c. *Secondly,* The various Durations of Silence, commonly called *Rests. Thirdly,* The various Degrees of Acuteness or Gravity in musical Sounds, as *A re, B mi,* &c.

Maxim II.

Strait Lines, whose Lengths are geometrically proportion'd to the various Durations of musical Sounds, will naturally and intelligibly represent those Durations. *Ex. gr.*

The first Line (being 2 Inches) represents a *Semibreve*.
The second is 1 Inch, and denotes a *Minim*.
The third is half an Inch, and signifies a *Crotchet*.
The fourth is a Quarter, and answers to a *Quaver*.
The fifth is an Eighth, and stands for a *Semiquaver*.

Maxim

Maxim III.

The Quantity of the blank Intervals, or Discontinuity of the Lines, will exactly represent the Duration of Silence or Rests. *Ex. gr.*

Maxim IV.

The different Degrees of musical Sounds, as *Gamut, A re, B mi,* &c. may be represented by the different Situations of those black Lines upon the red ones or faint ones. *Ex. gr. see* TAB. I. *Fig.* I.

Problem.

To make a Machine to write Music in the aforesaid Character as fast as it can be play'd upon the Organ or Harpsichord, to which the Machine is fixed.

Postulatum.

That a Cylinder may be made by the Application of a circulating, not a vibrating, *Pendulum*, to move equally upon its *Axis* the Quantity of 1 Inch in a Second of Time, which is about the Duration of a *Minim* in *Allegro's*;

Suppose the Cylinder *a* (see *Fig.* 2.) to be such, and to move under the Keys of an Organ, as *b, c, d,*

3   and

and nail Points under the Heads of the Keys, it is manifeſt, that, if an Organiſt play a *Minim* upon *c*, that is, if he preſs down *c* for the Space of a Second, the Nail will make a Scratch upon the Cylinder of 1 Inch in Length, which is my Mark for a *Minim*.

Again, if he reſt a *Crotchet*, that is, if he ceaſe playing for the Space of half a Second, the Cylinder will have moved under the Nails half an Inch without any Scratch; but if the Organiſt next preſſeth down *d* for the Space of half a Second, the Nail under *d* will make a Scratch upon the Cylinder half an Inch long, which is my Mark for a *Crotchet*. It will likewiſe be differently ſituated from the Scratch that was made by *c*, and conſequently diſtinguiſhed from it as much as the Notes now in Uſe are from one another by their different Situation in the Lines. (*Vide Fig.* 1.)

Theſe three Inſtances include all that can be performed upon an Organ, *&c.* (Maxim I.)

Therefore it is already demonſtrated, that whatever is play'd upon the Organ during one Revolution of the Cylinder *a* (*Fig.* 2.) will be inſcribed upon it in intelligible Characters. —— I proceed to ſhew how this Operation may be continued for a long time.

In *Fig.* 3. *aa*, *b*, *c*, *d*, are the ſame as in *Fig.* 2. Let *x* be a long Scroll of Paper wound upon ſuch a Cylinder as *z*. Let *eeee* be the ſame Scroll brought over the Cylinder *a a*, to be wound upon the Cylinder *yy*, as faſt as the Motion of *aa* (which is determined by a *Pendulum*) will permit.

It is manifeſt, that whatever is play'd upon the Organ during the winding up of *yy* will be written on the Scroll by the Pencils *b*, *c*, *d*, *&c.*

All

All the Graces in Muſic being only a ſwift Succeſſion of Sounds of minute Duration will be expreſſed by the Pencils by ſmall Hatches geometrically proportion'd to thoſe Durations. *Ex. gr.*

A ſingle Beat
A double Beat
A Shake
A Turn
A ſingle Backfall
A double Backfall
A Shake and Turn

If a Line commence exactly over or under the Termination of another, it is an Indication of a *Slur*; as

So a ſmall Interval indicates the contrary; as

Flat or ſharp Notes are implied by their Situation on the red Lines; the natural Notes being always drawn between them, *viz.* in the Spaces. (*Vide* Fig. 1.)

The Scroll may be prepared before-hand with red Lines to fall under their reſpective Pencils. It is the ſureſt Way to rule them after; tho' it is feaſible or poſſible to contrive that they may be ruled the ſame Inſtant the Muſic is writing.

The Places of the Bars may be noted by two ſupernumerary Pencils, with a Communication to the Hand or Foot of a Perſon beating Time.

Grave

III

Grave Muſic from briſk, ſlow from faſt, *&c.* will be better diſtinguiſhed by this Machine, than in the ordinary Way by the Words *Adagio, Allegro, Grave, Preſto,* &c. for, by theſe Words, we only know in general this muſt be ſlow or faſt, but not to what Degree, that being left to the Imagination of the Performer; but here I know exactly how many Notes muſt be play'd in a Second of Time; *viz.* as many as are contain'd in 1 Inch of the Scroll *per Poſtulatum* P. 447.

Laſtly, Whereas, in the ordinary Way of writing Muſic, you have either no Character for Graces, or ſuch as do not denote the Time and Manner of their Performance, here you have the minuteſt Particles of Sound that compoſe the moſt tranſient Graces mathematically delineated.

N. B. Tho', to facilitate the Demonſtration, I ſuppoſe the Pencils to be fixed under the Heads of the Keys, and conſequently to require a very broad Scroll to paſs under them; yet I intend the Pencils a more commodious Situation, *viz.* the Motion of the Keys to be communicated by ſmall Rods to them (which I know better how to do than to deſcribe, the Scheme would be ſo perplex'd). The Pencils are to be made of Steel, and ranged in cloſe Order like the Teeth of a ſmall Comb, ſo that a very narrow Scroll will do. I can prepare the Paper to receive a very black Impreſſion from the Pencils at ſo cheap a Rate, that, at the Expence of 6 *d.* in Paper, I can take in Writing all the Muſic that the ſwifteſt Hand ſhall be able to play in an Hour.

III.

BÜTTNER

A Dr. Büttner of Darmstadt invented a swinging sector instrument in which the platen struck upwards against the typeface upon depression of the selector.

CADMUS

Washington telegraphist Eugene Cadmus patented a typewriter which borrowed heavily from the printing telegraphs of the day. He used a sewing machine treadle instead of an electric motor to keep a pin barrel revolving, to be stopped by depressing a key on a piano keyboard. Printing by type wheel. The US patent was granted in 1872, which was late in the day for this design.

CALAHAN

E. A. Calahan of Brooklyn was granted a US patent in 1876 for a type wheel machine with index and indicator, printing on paper tape and inking by roller.

CANTELO

A small type sleeve machine with a three row keyboard and double shift was invented by J. L. Cantelo and patented in England in 1903. The machine was never manufactured but a single aluminum model, formerly in the author's possession, has survived. The type sleeve was mounted obliquely above the platen and located the desired letter by both turning and travelling along its axis before being brought down onto the paper. Inking was by roller.

From 1887 onwards, Cantelo filed more than ten typewriter patents which were abandoned, as well as several which were granted. These included one in 1894 for a musical typewriter to be attached to a piano keyboard and a braille machine in 1905, apart from the 1903 instrument described above.

CARAMASA

A machine by this name was reported to have been invented in 1885 but attempts at tracking down confirmation have failed. The instrument was said to have two piece type bars which pirouetted on their axes. Inking was by pad.

CARMONA

A type wheel instrument with a ninety-three character capacity operated by five keys and two levers was the subject of a number of patents granted from 1897 to Manuel Carmona of Mexico City.

CARPENTER

W. M. Carpenter of Missouri patented a stenographic machine in 1888 with sets of keys printing its own stenographic alphabet.

CARSALADE

Ten unsuccessful attempts at producing a syllable typewriter were made by the Belgian Paul Carsalade du Pont from 1905 to 1912. His fifth model is reported to have had a keyboard twenty inches high to accommodate the complex linkages.

CARTOGRAPH TESSARI

Typewriter Topics reported a machine by this name on display at the Office Equipment Exhibition in Venice but details are lacking.

CENTAUR

No details are available of this portable typewriter announced in London in 1925.

CENTURY

The Century Machine Co. of New York was assigned two patents granted in 1897 to the prolific Lee Burridge for one of his typically ingenious designs involving a keyboard of twenty-seven keys operating a mere nine key levers by a rocking motion employed on some of his other patents. These key levers activated a similar number of type bars which were rotated to the desired position, having three characters on each of three faces. The inking was by roller. Messrs. Hess and Stoughton subsequently patented some improvements on this basic design.

CERESETO

Vittorio Cereseto of Genoa invented a machine for the blind in 1907 with longer keys for the two thumbs—for spacing and carriage return—and a keyboard arrangement based on letter frequency. Exhibited in Venice in 1907.

CHAMBONNAUD

There were several versions of this French shorthand machine printing Roman characters designed to permit several keys to

106. Diplograph[83]

be pressed simultaneously. The instrument operated on the thrust principle and in fact looked like a small Wellington or Empire, but was not manufactured.

CHILDRESS
Patent granted to H. P. Childress of Memphis, Tennessee in 1902 for a cipher machine. Not manufactured.

CHRISTIANSEN
British patent granted in 1886 to B. F. Christiansen for an electric machine. Not manufactured.

CIENANT
This spring driven musical typewriter, which printed the score as it was being played, was the invention of Paul Cienant of Buffalo in 1902.

CLAVIGRAFO
An Italian shorthand machine, first called the Logomatografo, patented in 1872, with a separate piano keyboard for each hand and which printed on paper tape. Eight years later, the perfected machine was renamed Clavigrafo and was granted a patent in 1880, still with a piano keyboard but printing Roman characters. The project was dropped on the death of the inventor in 1884.

CLÉMENT
Abbé Clément was granted a French patent in 1857 for a remarkable machine which he had invented two years earlier. In the repressive post-revolutionary environment, it is hardly surprising that it landed him in hot water, for it enabled anyone to produce provocative pamphlets without needing to go to a printer. This was not the Abbé's intention of course, for he demonstrated his machine by typing out nothing more dangerous than the names of his pupils. The table-top device consisted of a number of alphabets placed end to end above the type plunger for each letter, with a manual selector and a pedal for bringing type and paper together. Dials at each end indicated the number and length of lines typed.

CLERK
US patent granted in 1890 for electrical typewriter.

CLEVENGER
Several patents were granted to this man who incorporated a $100,000 company in New Castle, Delaware, to manufacture

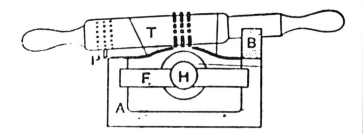

108. Thomas—1854 patent[50]

a $10 machine. His second effort was a telegraphic typewriter called Book Electric, manufactured in 1906, and his third was a 1911 patent for a plunger operated spherical index machine.

CODVELLE
Valentin Codvelle designed a machine featuring a horizontal circular housing containing keys in concentric circles. Printing on the flat bed was by down stroke and the housing travelled along a rack on either side of the base. Special keys for printing upper case by decorating the corresponding lower case letter. The French patent for this Daw and Tait ancestor was issued in 1861; a year later the inventor turned his attentions to a patent for a device designed to prevent trains from colliding and this sequence may well illustrate the fact that at the time railways were a surer bet than typewriters for fertile inventive minds. **(104)**

COLE
An 1884 patent was granted to J. W. Cole of Kansas for a primitive circular index machine.

CONRAD
An 1876 United States patent was granted to Hiram Conrad of Pennsylvania for a machine with a type on flexible steel strips fixed radially to a horizontal disk.

CONTI
An important contribution to the development of the typewriter was made by the Italian Pietro Conti whose Tachigrafo or Tachitipo achieved some degree of recognition in and after 1823. The machine itself has not survived, nor have drawings of it, but a detailed description in contemporary scientific journals point to a sophisticated up stroke design, the entire circular type basket of steel type bars moving a space at a time from left to right beneath a flat paper table which itself moved backwards or forwards for line spacing. A keyboard with a separate key for each character which the fingers could operate without moving the position of the two hands suggests a machine of roughly modern proportions which makes some later efforts distinctly grotesque. Other details are not explained: the type bar for instance apparently inked itself on the way to the paper.

Conti gave us the first type bar design, the first up stroke, the first keyboard, and the first design in which the paper remained stationary while the type moved, a revolutionary concept until very recently indeed. The Tachigrafo aroused considerable attention and several were made, although none has survived. Conti corresponded with his compatriot Giuseppe Ravizza for many years, and his death in 1856 overlapped the development and patenting of Ravizza's first model. It has been suggested that Ravizza's invention was modelled on Conti's and all indications are that this may well have been so.

COOKSON
A most attractive type plunger machine patented in 1885 by F. N. Cookson in which the type is located radially around a type wheel and thrust against the platen by a hammer. A handle to the right of the instrument geared to the type carrier is turned to select the letter on a circular index on which the letters are

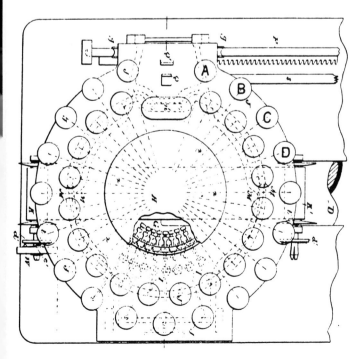

107. Lucas (US patent)

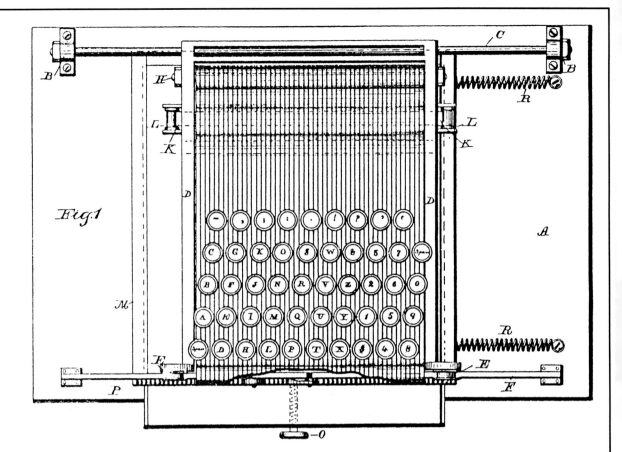

109. Worrall (US patent)

Fig.1

alphabetically arranged. The Cookson appears never to have been produced commercially.

COOPER

Several prototypes of a type wheel machine invented in 1856 by John Cooper have survived. The desired letter was selected from a horizontal circular index, as in a number of similar designs, but depression of a lever then rocked the carriage against the type, a design feature adopted some thirty years later by the manufacturers of the Peoples, Champion, and similar machines. Other progressive features included a carriage with platen and feed roller of distinctly modern design.

CORRADI

Piraino di Corradi of Palermo invented this machine for the blind in 1904 featuring six keys in two rows.

COUTTOLENC

A remarkable French patent granted in 1902 for a non-visible machine with the type wheel mounted under the platen.

COX C.

A certain Charles Cox of Brooklyn is reported to have perfected a machine capable of visible typing which 'should prove a best-seller if manufactured properly' (why, of course it should have!), after which nothing further was heard of the man or his machine.

COX J.

A primitive device built by John Cox about 1860—a vertical type wheel turned by a knob in the centre, with a single key above bringing type into contact with the paper which was positioned on a wooden strip on the base, without provision for locating it. The type wheel moved along a rod, the spacing was by ratchet. Its sole claim to fame is that it was a forerunner of several later cheap machines such as the Herrington and Dollar.

CRAIGMILES

Edwin Lee Craigmiles of Tennessee was granted a US patent in 1893 for a shorthand machine with four pairs of keys for the fingers of each hand, a pair for each thumb, and one for each palm. The keys followed the anatomy of the hand for easier operation, but nothing came of the design.

CRAM

The Cram Writing Machine Company of New York was to have introduced a combined writing and adding machine in 1907 but nothing ever appeared.

CRANDALL

An 1875 patent was granted to Lucien Stephen Crandall for an up stroke machine with a single curved row of eight keys and type bars with six characters on each bar. It was an abortive design, but Crandall's fame was assured by the later machines which bore his name.

CREED

The Reverend Creed would appear to deserve credit as the first man to have invented a writing machine. Martin mentions a man called Cred *(sic)* said to have invented a machine called Melograph in 1745.[63] Chapuis also refers to this device.[16] The Philosophical Transactions of the Royal Society call him Creed and his machine the Megalophone.[82] Neither Melograph nor Megalophone could be located in any other original sources and one must assume an error on the part of both Chapuis and Martin, although Grove's definitive Dictionary of Music and Musicians refers to Unger's invention as Melograph. However, the details of Creed's device are beyond dispute. These were published in 1747, two years before the first public claim by Unger, and must be given priority, especially as the 1747 reference, which includes a letter read to the Royal Society, already refers to the 'late' Rev. Creed (but without specifying the precise date of his death). We can speculate as to whether Unger and Creed developed identical machines independently of each other at the same time or whether one plagiarized the other. What we cannot dispute is that Creed communicated his invention before Unger, and by this act alone must be afforded priority.

Creed's writing machine consisted of a roll of paper advanced by clockwork at a precise rate of 'one Inch in a Second of Time, which is about the Duration of a Minim in Allegro's,' the clockwork to be regulated by means of 'the Application of a circulating, not vibrating, Pendulum.' A 'vibrating' pendulum might have proved more accurate, but he obviously specified a 'circulating' device, which might either have been a balance wheel or more probably a fly, for practical reasons in order to reduce the size. This clockwork movement turned a cylinder under the keys and paper passed over this cylinder in such a way that pressing a key caused the head of a nail to scratch the paper. The length of the mark obviously indicated the duration of the note, and parallel lines could be drawn on the paper before or after in order to read the score.

It may be argued that scratches on a piece of paper hardly constitute a sound basis for a writing machine—however, to make the scratches more visible, Creed considered that some form of permanent marking could be used ('I can prepare the paper to receive a black impression,' he states).

Since a roll of paper the entire width of a piano keyboard was hardly practical, Creed also proposed a system of rods and levers to reduce the width to 'a close order like the teeth of a comb, so that a very narrow scroll will do,' but gave no details of this arrangement 'which I know better how to do than to describe, the scheme would be so perplexed.' This is unlikely to fool anyone: every inventor in history would have given his eye-teeth for the opportunity of squirming his way out of mechanical trouble by claiming that certain elements of his invention were not too complex to design, but simply too com-plex to explain! However, in terms of Creed's rightful historical place this is largely irrelevant, for the cruder original version of his invention itself suffices to ensure his claim to being the first inventor of a writing machine.

Given the importance of this invention in the history of the writing machine, the corresponding entry (p. 445-450) in the Proceedings of The Royal Society is reproduced in its entirety.[82]

(25, 105 a-f)

DACTYGRAM
This tiny machine with five keys, any of which could be pressed simultaneously, was designed by a Frenchman called Georges Moulin in 1920.

DAMON
An American called Charles Damon was granted a patent in 1889 for a shorthand machine with one key per finger on either side of central keys for the thumbs.

DANEL-DUPLAN
A French patent was granted in 1864 for the Sténotype which was designed to be worn around the neck, with four keys for the fingers of each hand on the front of the machine and an additional key at the back for each thumb. The impression took the form of red and black dots and dashes, using the equivalent of a felt tip pen, on a roll of paper advanced by clockwork. It sounds an eminently sensible machine but regrettably, *'n'a donné aucun résultat pratique'* ('it offered no practical result').[27]

DAVIS
Type wheel patent granted to Edmund Davis of New York in 1877.

110. AEG 6 (Courtesy of Sandy Sellers)

111. Aktiv

DAVY
   The electro chemical telegraph was launched in 1838 by Edward Davy's discovery that certain chemical compounds would suffer decomposition and discoloration at the point of contact if subjected to an electrical impulse. Davy's discovery was the starting point for Bain's considerable work in this field.

DE DEDULIN
   French 1872 patent for a shorthand machine of radial type plunger design printing on paper tape through carbon paper.

DELTA
   One of fifty odd machines designed by Richard Uhlig, but in this instance not manufactured.

DEMENT
   Merritt H. Dement of Chicago was granted several patents from 1883 onwards for a machine in which type moved in grooves cut along the length of a cylinder.

DE MEY
   Another instrument in which type plungers were mounted on a horizontal wheel, patented by F. A. De Mey of New York in 1863.

DEMING
   Philander Deming of New York was granted a number of patents including two in 1876 for machines of up stroke design.

DESTAILLATS
   A type wheel design was the subject of a French patent granted to Abbé Destaillats in 1903.

DEVINCENZI
   Giuseppe Devincenzi invented an electric typewriter in 1854. His device had a keyboard mounted at the top of the machine above a comb—pressing a key down caused the corresponding tooth in the comb to bend. A cylinder with a row of spiral pins aligned with the teeth of the comb was kept turning by pumping a treadle. On pressing a key, the tooth caught the pin on the cylinder and stopped its rotation. At the end of the cylinder was a wheel covered with type plungers which stopped at the corresponding letter where an electro magnetic hammer banged it out against the paper. A small conveyor belt dipping into a trough brought ink to the type. Brett's comb and barrel, invented a few years earlier, provided at least part of Devincenzi's inspiration but the electro magnetic hammer was his own contribution.

DIAMOND
   This four row front stroke upright designed in 1923 was one of the abortive efforts of Sholes' son Fred.

DILLIÈS
   'Simplicity itself' was the Phonotype or Phonographe of Jean-Baptiste-Henri Dilliès, patented in 1866.[27] A straight line of twenty keys adorned the top of the machine, beneath which paper tape passed from one spool to another, with ribbon or carbon paper for inking. The design got beyond the patent stage according to contemporary sources, but never went into production.

DIPLOGRAPHE
   A table top machine invented by Ernest Recordon of Geneva, designed to print and at the same time to emboss the text, thereby permitting a blind operator to read what he or she was writing, was reported in 1877.[57, 83] The Diplographe used two concentric type wheels, a small one for printing and a larger one for embossing braille cells. A circular index was around the periphery on the face of the larger wheel. **(106)**

DODGE
   An 1885 US patent for a circumferential type plunger machine was granted to M. C. Dodge of Michigan.

DODSON
   A patent was issued to D. W. Dodson of Pennsylvania in 1884 for a multiple type wheel machine.

DOLD
   A shorthand machine printing up to ten syllables per line, using paper tape, invented in France in 1920.

DRAIS
   This flamboyant and prolific German inventor, best remembered for his contribution to the development of the bicycle

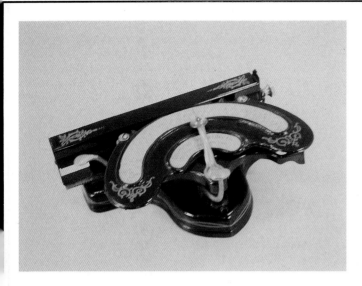

112. American (1) (Courtesy of This Olde Office, Cathedral City, California)

which bears his name, was also responsible in 1831 for a short-hand machine which marked Germany's entry into mainstream 19th century typewriter history. The instrument was housed in a wooden box roughly one cubic foot in size. Letters and combinations of letters of the alphabet, reduced to 16 elements, were recorded on paper tape advanced by clockwork. Despite claims to the contrary, Drais' machine was not the first to use a keyboard. Contemporary witnesses confirmed that the instrument was capable of amazing speeds which the inventor claimed exceeded 1,000 characters a minute! That kind of flamboyant extravagance was consistent with the inventor's well-documented character and, if at all creditable, must surely be related simply to the maximum number of random marks the machine was physically capable of making, using all fingers…and perhaps a few toes, for good measure.

DRIESSLEIN
Electric up stroke machine patented by C. L. Driesslein of Chicago in 1879.

DUJARDIN
The Tachygraphe invented by Antoine Dujardin in 1838 used a system of ink dots. A piano keyboard, with the letters in alphabetical order, was used to operate levers which converged to print what appeared to be random dots on paper tape until an ivory scale placed over the paper indicated the letter to which the dot corresponded. To save time and increase speed, it was suggested that syllables could be typed as arpeggios, but only, warned the inventor, if the component letters followed each other alphabetically (as in *cinq* or *mort*). Clearly, it was going to take a little time working this alphabetical conundrum out when taking a dictation—imagine trying to do this with polysyllabic words!—and therefore, despite further improvements using coloured dots to identify the code, the Tachygraphe was never produced.

DUKES
Harry S. Dukes of Missouri was granted several patents from 1897 onwards for a down stroke machine with a four row keyboard suitable for use either as a book typewriter, or, with an optional carriage, as a standard machine. The Dukes Typewriter Co. of Little Rock was to market the device in 1908, and a novel Vertical Plane Type Writer designed to print in books held upright in the machine, subject of a patent filed in 1900 and granted six years later, was also planned.

DUOGRAPHE
The 1907 invention of a clergyman called Stiltz and so called because pressing a key resulted in the simultaneous printing of a Roman character and the embossing of a braille cell.

DUPLOYÉ
An 1890 shorthand machine designed by Gustave Duployé of Paris, whose brother devised a well known stenographic system. It had a keyboard of fifteen keys and three bars, up to three of which could be pressed simultaneously.

ECKELS
Syllable typewriter of complex type wheel design with over eighty keys, patented from 1892 onwards by George Eckels of Chicago.

EDEL
Ernst Dettman and Erich Laude of Hamburg used their initials to name this 1922 thrust action machine which was not manufactured.

EDDY
Quite possibly the largest and most complex typewriter ever invented was the life's work of Oliver T. Eddy who has earned the dubious honour of being the first (but by no means the last) typewriter inventor to have been impoverished by his machine, the very complexities of which made its failure a foregone conclusion. His device, completed in 1850, consisted of seventy-eight keys in four rows and controlled a similar number of plungers which automatically inked themselves on their way to the flat paper carrier at the centre of the machine. A type carrier with six rows of thirteen characters each row had to be moved laterally and longitudinally each time a key was pressed. Spacing and inking was automatic.

EDISON
Thomas Edison patented an electric printing telegraph in 1872 in response to a commission from the Automatic Telegraph Co. As in so many of Edison's patents, the instrument had few if any strictly original features: a constantly rotating type wheel on a spirally-pinned shaft was controlled by keys on a two row keyboard using the established comb and pin barrel principle. The paper remained stationary as the type wheel moved along the line with printing by means of a hammer striking paper

113. American (2) (Courtesy of This Olde Office, Cathedral City, California)

against type wheel. Use of a continuous roll of paper permitted messages to be torn off when complete.

An electro-chemical telegraph attributed to Edison is also recorded.

ELEKTROGRAPH

An electric type plunger design with a flat paper table and a keyboard of some eighty keys was invented by Dr. Faber of Berlin in 1900. The plungers were electrically impulsed and apparently any number of clear carbon copies could be made by simply turning up the voltage as required.

ENNIS

An electrically operated machine with the type wheel placed vertically above the platen, patented by George Ennis of New York in 1901.

ESSICK

Teletype machine invented early in this century by a New Yorker of this name. Its applications included use for telegrams as well as news bulletins.

EXCELSIOR SCRIPT

Halstrick's interesting idea was taken up in 1899, but was never manufactured, by the Excelsior Script and Typewriting Machine Co. of San Francisco. It offered type cast from samples of the purchaser's own handwriting; or the script could be used for italicizing, with conventional type for the printing.

EXPRESS

Electrical contacts on an indicator and an index plate were used on this type wheel machine invented by Frenchman Jules Duval in 1900 but, despite auspicious beginnings, it was not manufactured.

FAIRBANK

J. B. Fairbank patented a phonetic printer in 1850, rivalling Eddy for the size and complexity of his design. The radial plunger machine sported a keyboard of forty-eight successive phonetic alphabets, removing the need for a movable paper carrier. The first letter of a line was selected from the first alphabet, the second from the second and so on. You left a space between words by simply missing out an alphabet altogether. A roll of paper tape was used and the whole printing surface was curved to accommodate the radial plungers. Inking by roller.

FARMER

L. C. Farmer of Minneapolis patented a swinging sector machine, similar to the World, in 1885.

FISHER

An ex-employee of the Oliver Typewriter Co. by this name was reported to have perfected a machine similar to the Oliver but with a standard four row keyboard and single shift. Promises of manufacture never came to anything.

FLAMM

Pierre Flamm's French patent dated 1863 for a mechanical compositor was the subject of an improved specification patented the following year in England. Details of keys, carriage and inking methods denote modifications intended to adapt the design for use as a typewriter.

FONTAINE

Two patents were granted to Louis Henri Fontaine in 1867 and 1869 for fine designs incorporating a number of exciting features. Type plungers were located around a circular carrier

114. American Visible (Courtesy of This Olde Office, Cathedral City, California)

115. Babycyl (Courtesy of Bernard Williams)

116. Bar Let (**l**), Underwood (**r**) (Courtesy of Bernard Williams)

turned by a handle and indicator above, depression of which caused the corresponding plunger to strike down against the paper by an electro-magnetic hammer. The machine printed around the platen which was turned the correct number of clicks by fine adjustment to the stop screw limiting travel of the selector lever, thus providing differential spacing. (A decade or so earlier Harger had a similar idea.) Four or more alphabets could be used simply by enlarging the size of the index and type carrier without increasing the selector slots which could be aligned with any alphabet required. The machine was never manufactured, and while its many features had already been individually invented and were to be used time and again in later production, the combination of all the features in this patent made for an interesting design.

FORBES

1876 patent for a stencilling machine similar to the Thurber Chirographer was granted to George Forbes in 1876.

FORTONI

Several models of musical typewriters were made by L. Fortoni of London. After twelve years of development, the first appeared in 1922. A modified version in 1926 was capable of typing 225 characters covering both music and alphabet, us-

ing a standard typewriter with a piano keyboard attached to the front.

FORTUNA

Hannoversche Apparatefabrik H. F. Tolle were to make an index machine of Hall-Kneist inspiration, but the 1908 announcement did not materialize.

FOUCAULD

Several designs of writing machines by Pierre Foucauld of France culminated in the Clavier Imprimeur of 1849, which was probably the first typewriter ever to be manufactured. Ten years earlier Foucauld had designed his Raphigraphe, which established the radial type plunger design he was later to perfect. Ten vertical plungers fanned out above the printing point and were pressed in chords to build up letters from their component horizontal, vertical, and oblique elements. Plungers at first perforated the paper, but were later fitted with pencil leads to make a visible impression.

The instrument was generally well received but the inventor dismissed it as unsatisfactory and began work on an improved version which was essentially one Raphigraphe upside down on another. The same message could be simultaneously printed on one page and perforated on another so that a blind opera-

117. Bar Lock (**r**), Commercial Visible 6 (**c**), Salter 6 (**l**) (Courtesy of Phillips, London)

tor could feel what he was typing while printing it for a sighted correspondent. The two machines were operated by a single set of ten keys in the middle and carbon paper was substituted for the pencil leads. Foucauld's continued unhappiness with this improved machine led him to design the Clavier Imprimeur. **(16)**

FRANCIS

This 1857 patent granted to Dr. Samuel Francis of New York for a Printing Machine was a fine concept of the up stroke class. The type bars were arranged in a semicircle and a piano keyboard was used. Neither warrant particular attention at this late date, however Francis did solve one problem which had plagued previous inventors—the linkage between key and type bar was slow, cumbersome, and prone to jamming. Francis' solution was to make the two elements independent of each other without permanent linkage. Pressing a key flicked the type bar up rather than pushing it, after which it fell back freely by gravity. He used a ribbon and made a simultaneous copy by inserting a sheet of thin paper between the type bar and ribbon. The copy then had to be held up to the light and read through the page.

Comment has been made about the similarity between the Francis machine and Giuseppe Ravizza's Cembalo Scrivano. There is a similarity, but it could equally be argued that similarities exist between Ravizza's and Conti's inventions. So many others came up with similar designs, including Sholes, that charges of plagiarism ought to be entertained right across the board, or not at all. **(39)**

FRANCUS

A Russian, Joseph Francus, was granted a French patent in 1904 for a machine with the type bars suspended above the paper.

FREY

Frederick Frey of Ohio invented an electric machine for the blind in 1900. It used four keys and a space bar and could be adapted for telegraphy, but was not manufactured.

GALLI

Celestino Galli designed a stenographic machine in 1830 which he called Potenografo but details are shrouded in mystery and it is not clear whether it ever left the drawing board. The design is known: it had a separate concentric circle of keys for each hand, one for vowels and the other for consonants, using radical plungers and printing on a strip of paper.

118. Bar Lock

119. Bar Lock 19

GALLOWAY

A precocious design of a linear index machine was patented in 1873 by John Galloway of New York. It was not manufactured, but several later machines, particularly the Sun and Odell, were of virtually identical design. Nine years later he patented a shorthand machine using a code of squares and dots, with all the keys operated by one hand.

GARRISON

G. C. Garrison was granted a US patent in 1885 for an indicator machine using a type sleeve and rectangular letter index.

GENSOUL

A Presse Sténographique patented in 1866 by a Frenchman called Henri Gensoul used three identical sets of eight keys to build up the components of a stenographic code consisting of the different elements of a square divided by two diagonals. Pressing the keys in chords produced the corresponding part of the square on paper advanced by clockwork. Despite some interest in the machine, the claims of the inventor were not substantiated even when he himself was at the keyboard and further development of the instrument for telegraphic purposes was equally unsuccessful.

GERMANIA

A German pneumatic machine the size and appearance of a small harmonium invented by a typing instructor called J. P. Moser. Compressed air was supplied by rubber balls which were squeezed when the keys were pressed. Due to be manufactured in 1900, the project did not materialize.

GERMANIA JEWETT

The company which imported Jewett components and assembled them in Germany produced the prototype of an electrified version of their machine but it was not manufactured.

GILMAN AND KEMPSTER

1884 patent granted to W. H. Gilman and D. E. Kempster of Boston, a year after a patent for a comparable device was granted to W. R. Perce of Rhode Island and there were several other similar designs, most notably those resulting in the manufacture of the Index Visible in 1901. Type wheel machine, with linear letter index; the type wheel was turned by means of tension on a cord attached to the indicator.

GLOSSOGRAFO

Mouth power was harnessed by the Italian engineer Amadeus Gentili in a number of futile efforts from 1882 onwards at recording the spoken word directly onto paper by breaking words down into their component sounds. Electrical contacts placed in different parts of the mouth and lips were wired to electro-magnets attached to pencils and were meant to produce lines on paper tape. It must be presumed that this is the same device, or at least a derivative, of the Glossograph proposed by a German called Züppinger in 1870.

GONOD

In 1827 a French librarian designed a piano keyboard machine consisting of twenty keys in four groups which printed a code of dots on paper tape thus inventing the shorthand typewriter. This was very early indeed and reflects the effect of the desire to type at the speed of speech on the development of the machine. It is also a fine example of the influence of the piano on early typewriter history, for apart from the keyboard itself, the shorthand system was based upon several keys being played simultaneously as chords. This idea recurs throughout the history of the machine. Paper tape was used but Gonod's suggestions for making the dots visible were clumsy and primitive even for those days: the keys struck the paper against a greased surface and the greasy dots were then made visible by sprinkling with coloured powder.

GRANT

A type plunger machine clamped to a table top and printing a shorthand code, called Monogrammic Alphabet, was patented in 1879 by H. M. Grant of New Jersey. The code was similar to that proposed by Benjamin Livermore some sixteen years earlier.

GRIFFEN

John Griffen of New York was granted a patent for a type wheel machine with a four row keyboard in 1902.

GRUNDY

A front stroke machine with a three row keyboard, offering fully visible typing, was the subject of a US patent filed by Arthur Grundy on 18th January 1887 and granted two years later. The inventor was just thirteen days too late to claim the honour of the first such design in the US, pipped at the post by Prouty and Hynes of Chicago whose invention had already been granted a British patent late the previous year.

GUILLEMOT

The inventive and prolific Adolphe Charles Guillemot patented a type wheel machine with a piano keyboard in 1859. The impression was achieved by means of a clockwork, weight, or pedal and inking by pad. An improved later model with three separate type wheels was no more successful than the first.

H

This initial is the only clue to the maker of an unusual machine in The Milwaukee Museum which they have tentatively attributed to E. E. Horton of Toronto, inventor of a type bar machine. The attribution seems difficult to justify, and it is this author's belief that H is more likely to refer to E. M. Hamilton of New York, whose small Automatic (also called Hamilton) bears some considerable resemblance to the H. Both are blind, type bar designs with unusual actions more easily comparable than those of the H and Horton. The Automatic however has a keyboard, while the H has a curved letter index.

HAEGELE-RITTER

This German patent dated 1880 was for a down stroke machine with a circular keyboard above a ring of type bars.

120. Bee
(Courtesy of Bernard Williams)

121. Bennett

HALL

Thomas Hall appeared on the typewriter scene in 1866 when his first patent was submitted, although he was reported to have been involved in developmental work for a decade or more before then. Unlike most of the inventors in those very early days, he did actually achieve fame and fortune later in the century—the machine to which he gave his name for which he is chiefly remembered is now prized by the modern collector.

His first invention was a failure despite being in every way a better design than his later successful index machine. One can only surmise that he was ahead of his time with his first effort for although it was heralded as one of the best in an almost contemporary report and otherwise reliably reported as capable of writing 400 letters a minute, the machine never actually reached production. Design was down stroke, the type bars in a circle above the flat paper table, with upper and lower case, ribbon, and differential spacing. The machine was successfully exhibited and orders were received but never filled. The machine which bore his name was patented in 1881 and his third design was called Century.

HALSTEAD

A downstroke machine with a semicircle of type bars at the rear and a piano keyboard at the front was patented as late as 1872 by the American Benton Halstead although he is reported to have completed the machine eight years previously. Type bars printed down on a flat paper carriage, weight driven, through a ribbon and spacing was by means of a stirrup.

HANSON

An oblique front stroke machine typing round a vertical platen raised a space at a time for each new line, this most unusual design was patented by Walter Hanson of Milwaukee in 1899 and partly assigned to Ole Lee who is reported to have continued working on the machine after Hanson's death. Sometimes called the Hanson-Lee.

HARDING

Two British patents were granted to G. P. Harding in 1870. Type plungers in a wheel or sector was one and a type wheel in which the paper struck the type was the other. Roller inking or carbon.

HARGER

An interesting design by Henry Harger was the subject of a US patent in 1858 and the ideas were used later in production machines. The type plungers were operated by a handle which advanced the paper as the plunger was being depressed. The type plungers of the wide letters were shorter than those of narrow letters, thereby altering the travel of the handle and giving full differential spacing.

HAUPTARGEL

A man by this name, from Leipzig, invented a machine for the blind capable of making eight copies at a time, which he was reported to have presented to a congress of blind teachers held in Germany. Date unknown.

HAZEN

Patent for a swinging sector design in a format strongly reminiscent of the Hall was reported to have been granted to Marshman Hazen of New York in 1901. Hazen's involvement with typewriters was extensive and continued until his death in 1909. He was for a time president of the Standard Typewriter Co., manufacturers of the Standard Folding. He was also a major shareholder of the Union Writing Machine Co. which produced the Brooks. A type wheel machine similar to the Peoples was another of his many designs.

HAZEN & UHLIG

Hazen teamed up with the prolific Richard Uhlig, who had more than fifty typewriter designs to his credit, in this 1908 patent for a swinging sector machine with four row keyboard.

HEISS

Electric swinging-sector patent granted to Frank Heiss of New York in 1892.

HERRINGTON

1881 patent granted to G. H. Herrington of Kansas for a type plunger design using a knob to select the desired character which was struck down onto the paper by means of a hammer.

HILGENBERG

US patent granted in 1882 to C. Hilgenberg of Illinois for an unusual circumferential type plunger design printing by up stroke and, thus, non-visible.

HITTER

An 1878 patent granted to a Jean A. Hitter, Jr. of Louisiana used type plungers in a horizontal wheel geared to a selector knob, with a separate knob for printing and roller inking.

HOCHFELD

A single machine capable of printing different alphabets or type up to a total of 180 was invented by Wilhelm Hochfeld of Hamburg in 1930. The type bars had four characters each, brought into correct register by two separate shift keys.

HOHLFELD

A report appeared in the records of the Berliner Akademie in 1773 of a musical writing machine invented some years earlier (1770 or 1771) by a man of this name (although Holfeld is also recorded).[16] This interesting device used a fan-shaped keyboard to record notes on a roll of paper advanced by clockwork. This is historically significant in its application during the 18th century of mechanical principles which were to become commonplace among 19th century inventors but was too late by several decades to warrant more serious consideration, having been preceded by Creed's comparable invention. **(26)**

HOOD

An early application of a vertical type wheel geared to a handle on a horizontal index, a concept which was to lead to the manufacture of a number of successful machines in later years, was the invention of a Scot called Peter Hood in 1857. Selection of a letter on the circular index turned the type wheel to the correct position, inking itself on a roller as it turned, whereupon it was lowered onto the paper clamped around a cylin-

122. Bing (Courtesy of This Olde Office, Cathedral City, California)

drical platen which was automatically advanced one space. A second machine is reported to have been built but details are lacking.

HOUSE

By 1865 when George House invented his type bar machine of up stroke design, these principles had already been applied by inventors time and again, but several details distinguish his efforts. He dispensed with the almost ubiquitous piano keys and substituted a keyboard of small cylindrical keys. Paper was clamped to a cylindrical platen and although it was a blind machine in that the typing was not immediately visible, the disadvantages of this design were minimized in that the machine typed around the platen so that the printing soon appeared. Inking was by roller. The machine was exhibited in Buffalo, New York, in 1885 but was never manufactured.

HÜLSEN

A patent was granted to a German by the name of E. von Hülsen in 1896 for a syllable typewriter but it was not produced.

HUTH

A. H. Huth of London was granted a patent in 1893 covering a front stroke machine with full keyboard.

JAACKSON

Jaan Jaackson's place is assured not so much by the typewriter he designed as by the reception it received. There are many examples of primitive inventions in typewriter history, but not even Cary's Writing Glove received such bad press. *The American Inventor* was apparently the first to report the device which consisted, we are told, of a hemisphere of curved ribs which carried the type. A handle in the middle of the sphere served to roll it in a concave inking pad whereupon it was simply turned till the desired letter was above the printing point.

'…a writing appliance absolutely without value…It is one of those impracticable inventions which are of no service to mankind and bring nothing to the inventor except loss of time and money.'[92] Take that, Mr. Jaackson!

JAQUET-DROZ

Completed in 1772, the writing automaton L'Écrivain by the famous Swiss clock maker Pierre Jaquet-Droz was not the first of its kind—von Knauss had earned that honour some twenty years earlier. But it is a tribute to his skill as a watchmaker of great renown that Jaquet-Droz was able to miniaturize the intricate mechanism of his machine and insert it in the body of a doll a mere thirty inches tall. It is still writing messages for crowds of admirers daily in the Neuchâtel Museum, more than 200 years after it was built.

Jaquet-Droz had more experience than most in the construction of such instruments. With associates Jean-Frédéric Leschot, Henri Maillardet, and others, he built Le Désignateur which drew sketches and portraits of famous people, including Louis XV, while a third played a keyboard instrument of harpsichord inspiration. There were many others which have not survived although records of them exist, including one on display at the Franklin Institute in Philadelphia which was damaged and subsequently rebuilt by Maillardet **(34)**. There are many examples of Jaquet-Droz's other creations, both large and small, in the form of musical and automated clocks and watches, music boxes, singing birds, and so on.

L'Écrivain was a marvel of complexity. Each movement of the doll was controlled by a separate linkage connected to its own lever, the movement of which was determined by its own cam 0.7mm thick in a vertical stack consisting of three sets of forty cams. Separate mechanisms control the movement of the arms for periodically dipping the pen in the ink, and shaking it over the inkwell to prevent blotting. All the above determine the movements which produce the given letter, while a separate unit controls a disk composed of forty different elements which, in succession, direct the above movement. For not only did the hand of the doll have to form the correct letters, but these had to be in the correct sequence, with the correct spaces between them, with inking and shaking of the pen, and with the doll's head and eyes turning to follow the written word. This surely makes a laughing stock of the suggestion that the typewriter could not be developed until the late 19th century because of the complexities of the machine.

Over the past two centuries Écrivain has travelled widely, composing different messages to suit the occasion. On its triumphant return to Switzerland early this century, for instance, after changing hands numerous times during its lifetime, the doll wrote *Nous ne quitterons plus notre pays* ('We will never leave our country'), after the First World War *Gloire à Joffre, Foch et Clemenceau* ('Glory to Joffre, Foch, and Clemenceau'), and during the depression *Pensons à tous les chômeurs de notre pays* ('Let us think of all the unemployed in our country'). More recently, and more prosaically, during an overseas visit complete with armed guard in armoured transport, it was reduced to the ignominy of writing *Welcome to the Swiss Watch Festival*.[32] Yet another equally spectacular writing automaton is to be found in the Beijing museum in China **(31, 32, 33)**. Fitted to the base of an 18th century clock signed by the Lon-

don maker Timothy Williamson, this *frère puisnè* of l'Écrivain is almost certainly a Jaquet Droz original adapted either in London or quite possibly in China itself which may help to explain some unresolved mysteries involving Charles Paris.[18]

JEFFREYS & EDWARDS

Linear index design for which a British patent was granted to Messrs Jeffreys and Edwards of London in 1894.

JOHNSON

A US patent was granted to Charles Johnson in 1869 for a Mechanical Typographer.

JOHNSON BOOK TYPEWRITER

J. W. Johnson was granted a US patent for a book typewriter in 1899, the marketing company to exploit the patent had already been formed three years earlier, but the project did not materialize.

JUNDT

John E. Molle, a Wisconsin watchmaker who later put his name on a front stroke machine of limited success, began his career in typewriters with an instrument of similar inspiration which was to have been marketed with the help of a man called Jundt and called after him, but the project did not materialize.

KAHN

A 1903 French patent was granted to Kahn for a shorthand machine which printed a shorthand script.

KALEY

J. A. Kaley of Ohio was granted a patent in 1885 for a small machine with type at the ends of flexible teeth cut into the perimeter of a disk using pad inking.

KENT

An electric typewriter with a vertical type wheel moving along a horizontal shaft parallel to the stationary platen was announced in 1892 by the Kent Writing Machine Co. A share issue was announced but this interesting project never materialized.

KITZMILLER

A front stroke type bar machine patented in 1904 by George Kitzmiller of Virginia and assigned to the Electric Typewriter Co.

KLACZKO

A prototype three row keyboard bilingual machine for Russian and German with five characters per type bar was designed in Germany in 1913 by Max Klaczko, who for over a decade had been performing multi-lingual modifications to

123. Blickensderfer 8

124. Blickensderfer Electric (Courtesy of the late Paul Lippman)

existing machines such as Ideal, Commercial, Minerva, and Regina.

von KNAUSS

If proof were needed to refute the contention that the typewriter was essentially a late 19th century phenomenon because mankind lacked the mechanical ability to produce it earlier, then one need look back no further than Friedrich von Knauss. For although, strictly speaking, this eminent maker produced writing automata rather than the species of writing machine we are dealing with, no genealogy of the typewriter would be complete without a look at the remarkable achievements of this inventive mechanical genius.

Von Knauss produced his first automaton in 1753 after dedicating some twenty years to perfecting it, and it was displayed in France amid universal admiration. Five years later, a second model was made for the Viennese court—it is still in Vienna—while a third, built the following year, went to the Grand Duke of Florence. The message it wrote was merely *Huic Domui Deus Nec Tempora Ponat*. The most complex however, brought the species to perfection in 1760. It had progressed from writing a mere five words at a snail's pace to composing a much longer message more quickly, inking its quill pen automatically every few words as it wrote.

Monsieur, Tant parfait, que je fut il y a vingt ans, suis-je même encore à cette heure par le grand génie de mon inventeur, et toujours prêt à tous présents, d'écrire tout ce qu'ils puissent désirer. Glorifions donc la Providence de corps et d'esprit, pour qu'elle bénisse le bon dessein de mon créateur, d'oser faire publier au grand monde, mon art, et mes utiles services, par une description générale imprimée, afin que l'on sache, que

je sois le premier parfait au monde de cette nature, et que je sois encore, malgré tous ses envieux, toujours Monsieur le même plus fidel Secrétaire. Vienne, le 15 Octobr.[20] (See Chapter One for a translation of this text.)

One might reasonably assume that if he was capable of the above, von Knauss would have had no difficulty in building an ordinary typewriter but clearly chose not to do so because there was nothing in it for him. He made automata, and was filling a demand for a new product. Dukes and kings were his patrons, and as their writing was done for them by scribes and secretaries, they had no need of typewriters. **(27, 28, 29)**

KNICKERBOCKER

The Knickerbocker Typewriter Company with a factory in Connecticut is known to have changed ownership in 1912 and after reorganization was to become the Defiance Typewriter Co. but nothing further was heard of the undertaking.

KOCHENDÖRFER

One of the early attempts at a practical electric typewriter was a battery operated machine with a three row keyboard patented in 1901 by Karl Heinrich Kochendörfer, the Leipzig based sewing machine manufacturer who had already produced typewriters called Eureka and Imperial. The project did not proceed beyond the drawing board.

KOHL

A modified version of the Hansen Writing Ball was made in the 1880s by the Dane Alexis Kohl as a monalphabetic substitution cipher machine. Called Cryptographe, the type plunger machine printed on paper tape like Hansen's early telegraphic model. Several improved prototypes were made but without success.

125. Blickensderfer Electric (Courtesy of Tom Fitzgerald)

PRICE LIST OF PARTS
OF THE
BLICKENSDERFER TYPEWRITERS

	Type wheels, each	$1.50
	Ink rolls, any color, per doz.	.75
501	No. 5 Main Frame,	5.00
501 B	No. 7 "	5.00
502	No. 5 Top frame,	2.50
502 C	No. 7 " large,	2.50
503	Sleeve,	.75
504	Right sector	.50
505	Left sector	.50
506	No. 5 Space Key	.50
506 B	No. 7 Space bar	1.50
507	No. 5 Carriage frame,	1.50
507 B	No. 7 "	1.50
507 C	12 in. "	1.75
507 D	14 in. "	2.00
507 E	18 in. "	3.00
508 B	Attaching bar,	
510	Long bail,	
511	Right bail,	
512	Left bail,	
513	Main shaft washer,	.05
515	Middle " "	.25
515 B	Key button attaching spring,	.05
516	Lower bank key levers, each	.25
517	No. 5 Space lever,	.25
517 B	No. 7 "	.25
518	Space lever cam,	.15
519	Capital shift lever,	.25
520	Figure "	.25
521	Lifting lever No. 1,	.25
522	" " 2,	.25
523	Shift lever lock plate,	.05
524	Front comb,	.50
525	Upper comb,	
526	Key lever holding plate,	.20
526 B	Carriage holding plate,	.05
527	Space plate,	
528	Spacing standard,	.10
529 C	Feed pawl guide plate	.05

531 B	Feed pawl stop plate,	.05
532 B	Automatic lock frame friction spring,	.05
532 C	Sprocket wheel,	.50
533 C	Carrier,	.50
533 A	"	.50
534 D	Snail shield,	.10
535 B	Snail,	.50
536	Type wheel shaft lifting plate,	.05
536 B	Main spring adjustment lever,	.05
538	Type wheel shaft lock plate,	.05
539 D	One piece front stop,	
541 C	Back stop bail,	
542 B	Back stop,	.15
543	" bail holding plate,	.05
543 B	Carriage roller frame,	.25
543 C	Extension roller frame,	.25
544 B	Carriage rack,	.50
544 C	12 in. "	.50
544 D	14 in. "	.60
544 E	18 in "	.75
546 B	Type wheel lock catch,	.05
547	No. 5 Ink frame,	.25
547 B	No. 7 "	.25
548	Ink frame latch,	.05
549	No. 5 Pointer,	.15
549-5	Lock pawl,	.25
550	Bell trip spring,	.10
550 C	Receding pointer,	.25
551-5	No. 5 Right carriage standard,	.25
551-7	No. 7, "	.25
551 C	Receding pointer frame,	.15
552-5	No. 5 Left carriage standard,	.20
552-7	No. 7 "	.20
552 B	Automatic lock arm,	.10
553 B	No. 5 Paper shield,	.20
553-7	No. 7,	.20
554 B	Scale supports, (each)	.15
	Pressure roll arm, (each)	.15
554 B	Elevated Scale,	.25
554 C	12 in.	.10
554 D	14 in.	.15
554 E	18 in "	.50
555	No. 5 Name plate,	
555 B	No. 7	.10
	No. 5 Line space scale,	

558	Line space plate,	.15
559	Ratchet wheel,	.35
560	No. 5 Bell,	.25
560 C	No. 7 Bell,	.35
561	No. 5 Bell striker,	.10
562	No. 5 " spring,	.05
563	Carriage stop plate,	.10
565	Carriage washer, thick,	.05
565 C	" thin,	.05
566 B	Feed pawl and link, assembled,	.35
567	Spring protection plate,	.20
568	Lower paper guide,	.05
569	No. 5 Carriage scale,	.35
569 B	Paper plate,	.35
569 C	Black finish scale,	.40
569 F	No. 5 Paper guide arm,	.50
570 F	No. 7 "	.50
571	Ink frame holder,	
571 B	Elevated scale pointer,	.05
572	Automatic lock frame,	.10
573	No. 5 Spring tension lever,	.10
573 N	Automatic lock plate,	.10
574	No. 5 Carriage knurl,	.25
574 B	No. 7 " Right and Left,	.35
575	Bail shaft,	
579	Space lever washer,	.05
576	Key buttons, each,	.15
576	" made to order,	.25
580	Space lever cam stud,	.05
581	Space plate roller No. 1,	.05
581	Space plate spring stud,	.05
582	Sprocket stop,	.10
582	Bell trip,	
583	Sprocket stop base,	.20
589	Space key pin,	.05
591 C	Main action shaft,	.35
592 D	Main action journal,	.05
593	Type wheel shaft,	.35
594	" nut,	.05
597 C	Pinion,	.75
598	Pinion key,	.05
599	Pinion and attaching bar nut,	.05
601	Name plate rivet,	.05
601 B	Patent plate rivet,	.05
602	No. 5 Main spring stud,	.05
602 D	No. 7 " action spring stud,	.05
603 C	Paper guide thumb screw,	.10

605	Steel assembling nut, .1,	.05
607	Automatic pointer stud,	.05
610	Ink frame rivet,	.05
611	Ink frame pin,	.05
612 C	Special shift lever washer,	.05
615	Main spring adj. lock nut,	.05
615 B	Line space "	.05
615 C	" for two-piece door case,	.05
616	Carriage rod nut, nickel, .125"	.05
617	Rubber bail,	.25
617 C	12 in. "	.30
617 D	14 "	.35
617 E	18 "	.35
618	Bell nut,	.05
620	Line space adj. screw,	.05
620 B	Main spring and tension lever adjustment screw,	.05
622	No. 5 Carriage rod,	.15
622 B	No. 7	.20
622 C	12 in.	.20
622 D	14 "	.25
622 E	18 "	.30
623	No. 5 Tension arm adjustment screw,	.05
625	Pressure roll arm pin,	.05
626	No. 5 Bell striker pin,	.05
627	No. 5 Margin thumb screw,	.10
628	No. 5 Paper rack,	.20
628 B	No. 7 "	.20
639 C	Snail washer, concave,	.05
639 D	.035" Pinion washer, flat,	.05
640	Shift lever washer,	.05
641	Front stop spring stud,	.05
649 D	Main action shaft collar, R & L,	.10
650	Key lever spring,	.05
651	No. 5 Space lever spring,	.05
651 C	No. 7 "	.05
652	Lifting lever No. 1 spring,	
653	No. 5 Space plate spring,	.10
653 B	No. 7 "	.10
654	Feed pawl spring,	.05
656	Type wheel shaft spring,	.05
657	Action lock plate spring,	.05
658	Front stop spring,	.05
659 B	Lock pawl spring,	.05
660	Ink frame spring,	.05

3   4

126. Blickensderfer 'Price List of Parts'

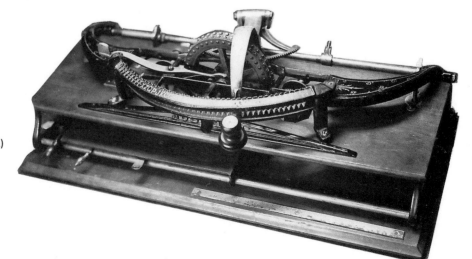

127. Boston (Courtesy of Tom Fitzgerald)

128. Burns (Courtesy of Tom Fitzgerald)

KROMAROGRAPH

A musical typewriter connected to the keyboard of a piano which recorded the notes played by means of longer or shorter strokes was invented by Lorenz Kromar of Vienna in 1904, roughly a century and a half too late.

LABRUNIE DE NERVAL

An instrument…which was not quite a typewriter but not exactly a printing press…was the Stéréographe, patented in 1844 by the French writer, traveller, and eccentric Gérard Labrunie (de Nerval). It consisted of a series of identical type wheels mounted on a spindle. Each wheel had two alphabets around its periphery—the lower one in relief, the counterpart engraved. A small lever next to each type wheel served to turn it to the desired letter, thus forming a line of text, whereupon two levers acting upon the spindle were pressed down onto the page held flat beneath the type wheels, having previously been inked by a roller. A small crank on a pinion underneath was turned to advance the paper to the next line.

The classification of this instrument is worth considering, since Labrunie's type wheels were used to print a line at time, instead of a letter at a time. Strictly speaking, we distinguish between typewriters and printing presses precisely on this fine point. If Labrunie had designed a machine with exactly the same specifications but with a single type wheel only, there would have been no disputing its classification as a typewriter, and examples of just such primitive designs were mass produced by different manufacturers later in the century. It is ironic that, considered from our microcosmic viewpoint, Labrunie risks disqualification from the ranks of typewriter inventors by improving upon what is indisputably a typewriter. Joseph Bramah's comparable invention presents a similar dilemma.

LAMONICA

Several stenographic machines were built by an Italian called Luigi Lamonica from 1867 onwards and apparently enjoyed a degree of success but were never manufactured. Early models were large and complex, consisting of some 5,000 individual pieces, but were later simplified and reduced in size.

The existence of the machine is well documented but details of the design are not recorded.

LANARI

Enea Lanari of Italy invented a shorthand machine with twenty-three keys in three rows in 1905 but it was not manufactured.

LARSON

S. P. Larson of Chicago was granted a patent in 1895 for a device consisting of a vertical type wheel riding on rods above a cylindrical platen. It was not manufactured.

LASAR

A large and heavy machine of down stroke design similar to the Horton is reported to have been manufactured in considerable numbers somewhere around 1898 to 1900,[92] although others report that only a few experimental examples were made.[62] The inventor was G. H. Lasar of Missouri who held over twenty patents dating back to 1889. A man called P. K. Lawrence was reported to have had an interest in marketing the machine, but history is far more likely to remember him for contributing his initials to Wrigley's P. K. chewing gum. Patent infringements were apparently responsible for the project's early demise—the typewriter's, that is, not the chewing gum's.

LECHET

A sole reference to a printing device for the blind invented by Claude Lechet appeared in a short news item journal in 1822. Details lacking.[26]

LEGATO

Oskar Fischer of Berlin invented this electric musical type wheel device in 1912 but manufacturing plans did not materialize.

LEGGATT

A portable machine of thrust design with a three row keyboard and double shift was marketed under the name Rochester by Rochester Industries Inc. in 1923, a year after the company had acquired the assets of the Leggatt Portable Typewriter Corporation. The machine was never marketed under the name of the company which originally developed it.

LEMING

A. G. Leming of Arkansas patented an unusual design in 1883 featuring a full circle of keys above a corresponding circle of type bars of down stroke action.

LE ROY

'Mr Sang relates in Engineering Magazine that one day, while looking through the files of the Journal des Savants, he came across an account of a machine presented to the Academy of Sciences of Paris, during the early years of the 18th century. Mr Sang remembers that it had individual type bars and was the work of Leroy of Versailles who was styled Horloger du Roi (Watchmaker to Louis XIV).'[91]

The Le Roys were a very famous family of clock makers and a type bar machine dating from the early 18th century by a technician of Le Roy's calibre would surely have turned typewriter history on its head. Regrettably, however, this information has proved impossible to verify. Despite the fact that this is a well documented family, no record of any attempt at building a writing machine has been found and, even more disturbingly, on my behalf a researcher has combed the Journal des Savants of the Académie des Sciences with great care over a period exceeding the whole of Le Roy's lifetime, not just the early 18th century, and the article Mr. Sang mentions is nowhere to be found.

LINDGREN

The platen rocked forward and struck the type wheel, instead of the other way around, on this design patented in 1881 by J. F. Lindgren of Illinois, ten years before this feature was popularized on the Peoples etc. A handle on top of the device was turned for letter selection and pressed down for printing, with slots for alignment. Inking was by roller.

LITTLEDALE

A gentleman of this name was the inventor of a typewriter for the blind which, by the insertion of carbon paper, could be used to provide a visible imprint. The device was exhibited at a meeting of the British Association in York in 1844 by the Rev Taylor and a description was published in the reports of the meeting. The machine consisted of a large wooden box containing a row of type and the means by which individual letters were brought beneath a hammer, with the paper in a frame advanced a space at a time at each hammer stroke. Metal type was also suggested.

LIVERMORE

The Hand Printing Device or Mechanical Typographer patented by Benjamin Livermore in 1863 was an interesting effort at producing a machine which could print twenty-six different characters approximating to the letters of the alphabet, using a mere six keys printing the four sides of a rectangle and the two halves of the bisected diagonals. He was not the first (nor the last) inventor to build up letters or symbols in this manner but his machine was small enough to be held in the hand (the name obviously relates to this feature). Up to four keys were pressed simultaneously to build up the more complex letters of the alphabet.

LIVOCK & HERMANN

An unusual design reported by Budan, but not confirmed elsewhere, this American machine featured a flat paper carriage inclined obliquely above a circular type basket.[12] Above these was an arm at the end of which was an inking pad, facing downwards. The type bar struck the paper against the pad, leaving the impression visible on the upper side of the paper. This is the only oblique up stroke machine ever designed. Mares also lists it.[62]

LOTTERHAND

A machine by this name featuring duplicating facilities was due to be manufactured by the Wales Adding Machine Co. of Wilkesbarre, Pennsylvania, in 1906 but nothing further was heard of the project.

LUCAS

George Lucas of New York patented a down stroke machine with circular keyboard in 1885, more than a year after Thomas and Hilder Daw patented a comparable design on the other side of the Atlantic. On the Lucas design, the paper remained stationary while the circular type bar and keyboard assembly travelled a space at a time along a rack. A platen and feed rollers were used; ribbon for the impression. **(107)**

MACKNESS

Shift key machine by a Scot called C. F. Mackness dated 1895. No further details.[62]

MAGNETOGRAPHE

Eugène de Neufbourg was granted a French patent for an electric typewriter by this name in 1877.

MAHRON

Type sleeve machine without further details.[62] Not manufactured.

MAILLARDET

A writing automaton attributed to Henri Maillardet, an associate of Jaquet-Droz, is in the Franklin Institute in Philadelphia and owes its attribution to the message the doll itself wrote after it had been restored. It is now believed to have been left unfinished by Jaquet-Droz and later completed by Maillardet who toured with it in the early 19th century, but this is not certain and there are those who believe that Maillardet was responsible for the device on his own, having previously been involved in one commissioned by George III for the Emperor of China in 1792.

Maillardet's automaton was perhaps less sophisticated than the others by Jaquet-Droz, who enclosed the entire mechanism within the body of the doll while Maillardet's machinery was located in a large case which formed the base of the instrument. Other than that the mechanical principles are the same, the movements of the doll controlled by cams turned by clockwork. On the other hand, by sacrificing miniaturization, Maillardet was able to enlarge the repertoire of the device to exceed that of any of its competitors, with a range of poems and drawings dwarfing their achievements.

Maillardet together with an impresario called Philipsthal toured extensively with his automaton delighting huge crowds wherever he went. After his death however it fell on hard times. It is believed to have toured with various companies, finishing in the hands of P. T. Barnum one of whose museums, in Philadelphia, is known to have contained automata at the time it was destroyed by fire in 1851. What happened after that is not known but in 1928 the present automaton, badly ravaged by fire, was delivered to the Franklin Institute and restored some years later, whereupon it revealed its identity by composing a couplet in French, followed by the words: '*Written by the Automaton of Maillardet.*' Regrettably, some liberties were taken during the restoration and the ball-point pen the automaton now uses is the result of the jettisoning of that part of the movement controlling the inking of the original quill pen. Almost as serious, from a purist's point of view, is that the original soldier boy is now either a transvestite or a trans-sexual, which represents a piece of licence the museum might do well to correct. **(34)**

MARCHESI

The London Exhibition of 1851 awarded a medal to the Italian B. G. Marchesi for a device for the blind which he had invented the previous year. Capable of printing embossed or visible characters, it was used for a time by the Blind Institute in Milan but was not adopted outside Italy.

129. Burroughs (Courtesy of This Olde Office, Cathedral City, California)

130. Burroughs (Courtesy of Tom Fitzgerald)

MARRIOTT
1904 US patent granted to J. H. W. Marriott of Maryland for an electrically assisted down stroke book typewriter.

MARTIN (1)
The London International Exhibition of 1862 recorded a machine for the blind submitted by John Martin which is now in the Science Museum Collection. It consists of a horizontal wheel carried on a frame above a flat paper table with vertical type plungers around the circumference of the wheel which are brought into contact with the paper by means of a handle.

MARTIN (2)
An American patent was granted in 1882 to a J. Martin of Ohio for a circular index machine printing on a flat paper table.

MAULER
A typewriter for the blind, reported and illustrated in 1887 in a German periodical and attributed to a man by this name, consisted of a horizontal type wheel with Braille cells towards the circumference and their corresponding characters towards the centre. Paper in a frame pivoted above the type wheel was lowered onto it by means of a handle.

McCALL
An automatic typing device based on perforated paper of player piano inspiration and also applied to automatic telegraphs was to have been manufactured by the McCall Automatic Typewriter Co. from 1906, with the patent dated two years later. The inventor was T. A. McCall of Columbus, Ohio.

McKINLEY
A design which featured a horizontal indicator geared to an obliquely mounted type wheel was patented in 1880 by Carl McKinley of Washington. Typing was produced by pressing down on the indicator, causing a hammer to knock the paper against the type wheel.

McKITTRICK
G. McKittrick of New York was granted a patent in 1881 covering a machine with a two row keyboard controlling a rotating pin barrel.

McLAUGHLIN
Numerous patents were granted from 1887 onwards to J. F. McLaughlin of Philadelphia for an electric machine of up stroke design.

MEGAGRAPH
Charles Augustus McCann was granted a British patent in 1898 for an enormous typewriter designed for typing texts of billboard or poster proportions onto paper. It had upper and lower case, differential spacing, and inking by roller.

MELOGRAPHE
A Brazilian priest called Jose Joachim Lucas was responsible for this musical typewriter, first reported in 1925.

MERGIER
An unusual application of the chord principle, in which the typist pressed any number of keys simultaneously, was patented in 1892 by a Frenchman called Guillaume Emile Mergier. The machine picked out the components of each chord and typed them consecutively but only did so in alphabetical order, so that by the time the operator had sorted out when he could and could not type a chord, depending on the order of the letters in the word or syllable he wanted to type, he would have lost more time than he could ever have gained.

MIDGET
Harry Bates of Chicago filed a patent for a machine by this name in 1920 (granted 1924) and Wellington P. Kidder was said to have perfected it into the Rochester of 1923.

MILL
The honour of the world's first typewriter patent goes to a British engineer by the name of Henry Mill who was granted patent 395 in 1714 for '*an artificial machine or method for the impressing or transcribing of letters singly or progressively one after another, as in writing, whereby all writings whatsoever may be engrossed in paper or parchment so neat and exact as not to be distinguished from print...*'
This ought to have qualified Mill as the undisputed inventor of the typewriter about a century or so before his time, but unfortunately this is not the case, for not only is there no record of the device ever having been made, but neither have details of the design itself survived. In Mill's day patents were granted without specifications having to be supplied, a fortuitous courtesy to inventors which was later to be denied them. Speculation regarding the patent is not lacking, however, and it has even been suggested that Mill's job as chief engineer of the New River Water Company would have left him no time in

which to make a model of his invention, although it seems improbable that he would have needed to dedicate his entire life, around the clock, to his aquatic duties. What is infinitely more likely is that Mill did not make his 'artificial machine' or alternatively supply details of its construction because he was not by law required to do so. In assessing typewriter history one suspects that many later inventors would have given just about anything to avoid having to provide proof that their inventions could actually be made, and made to work.

MINIMAX

There appears to be a single source of information about this machine said to resemble the Ideal and to date from 1907.[91]

MITTERHOFER

Peter Mitterhofer was at one time promoted as *the* inventor of the typewriter, his contribution to the history of the machine exaggerated to enhance and embellish what should otherwise have been a perfectly straightforward matter of record. Mitterhofer was born in that part of Northern Italy which is more Austrian than Italian and his arrival on the scene is relatively late in terms of the history of the typewriter: he was more or less a contemporary of Sholes in the United States. There are parallels to be drawn between the two inventors and even between the way they both began, but what Mitterhofer lacked, perhaps more than anything else, was a James Densmore in his life—a man capable of driving the invention out of the stuff of dreams and into the hard world of reality.

Mitterhofer was never able to get his crude, hand-whittled prototypes into marketable form—he would have needed an outfit comparable to Remington's to have achieved that. His four experimental machines, seen together, present a picture of a man grappling forever with problems he was unable to solve. The machines were all type bar instruments, up stroke, with the type bars arranged in a basket—a well established design by that time. They shared that feature with the Sholes and Glidden, and there are further parallels in the design of the leverage between keyboard and type bar. The first Mitterhofer model printed only upper case, as did its US counterpart, but on his third model he provided upper and lower case by means of a shift mechanism, only to abandon it altogether in favour of a double keyboard on his last machine.

What to do with the paper was also a problem which he never managed to resolve successfully. On the first model he tried a flat paper table, the second model was left incomplete in this respect, while on his last two models he used a cylindrical platen on a worm, typing around rather than along the paper. Inking was also left largely unresolved: he by-passed it altogether on his first machine by providing needle points to perforate the paper; it is missing from his third model (there is of course the possibility that he had temporarily abandoned the idea of inking in favour, perhaps, of carbon paper) while on his last machine, inking was performed by a brush. **(24)**

MODEL

There were two efforts at producing a machine with this name, neither of which was successful. The first was in 1909 when the Model Typewriter Co. of Harrisburg planned to produce a sixty dollar machine but went out of business four years later. The second was one of the many designs of Richard Uhlig which he is reported to have been perfecting in the early 1920s but which never left the drawing board.

MODERN

Another of the many designs of the prolific Richard Uhlig was this 1923 effort which never left the drawing board.

MOLITOR

The notoriously inaccurate Budan lists a machine by this name under those using a single key, without specifying whether it was merely an alternative name for an existing instrument or else a separate design which was never marketed.[12] (Budan was so cavalier with orthography that one is tempted to speculate whether the name ought not to have been 'Monitor,' but there is as yet no evidence to substantiate this.)

MOON HOPKINS

Hubert Hopkins of St. Louis combined an up stroke typewriter with an adding machine in 1902 but the product met with little commercial success and changed hands several times before Burroughs Adding Machine Co. bought it up in 1921.

MOORE

An 1876 patent for a type wheel device was granted to Charles Moore of West Virginia, a quarter interest in which was assigned to a shorthand reporter called James Ogilvie Clephane who was better known for having conducted some tests and trials on early Sholes and Glidden machines, which performed so poorly that it may well have prompted him to turn his thoughts away from type bars in favour of the type wheel.

MORGAN

Numerous patents were granted to George Morgan of Ohio from 1876 for machines using type wheels and hammers for printing in up to three colours using a roller for inking. The Typographical Machine Printing Company of Washington was to undertake production but the project did not materialise.

MORSE

Samuel Morse is not a man normally associated with typewriter history and his contribution to other fields needs no introduction. What is generally less known is that he was the first inventor to come up with a permanent telegraphic record. His 1835 invention, in which the receiver embossed the signal, was the forerunner of the printing telegraph. Morse persevered with the device for a few years, using various different inks and inking systems, only to drop the idea in favour of the buzzer which produced its distinctive audible signals.

MOSSBERG

A machine by this name is said to have existed although no details are available other than the possibly related fact that the Granville Automatic was produced by a company called Mossberg and Granville Manufacturing Co.[92]

131. Caligraph 3 (Courtesy of Christie's South Kensington, London)

132. Canadian Scout (Courtesy of Sandy Sellers)

MÜLLER

A syllable machine said to resemble a Hammond but of type wheel design was patented in 1902 in Germany by F. J. Müller.

MULLIGAN

B. T. Mulligan of New York patented a linear index device in 1885 in which printing was achieved by means of a hammer striking from the rear.

MURPHY

J. R. Murphy of Pennsylvania was granted a patent in 1885 for a blind up stroke machine of Remington inspiration.

MUSIGRAPHE

A Belgian company called Allman & Cie were granted a patent for a musical typewriter in 1908 to be called Musigraphe but the project did not materialize.

MYERS

Several patents were granted to an Englishman called F. Myers but only one of the designs was manufactured briefly as the

Mercury. Apart from that, the inventor was responsible for a down stroke machine patented in 1886 in which keys were attached directly to type bars arranged in two semicircles, front and rear, above a central platen.

Myers was also responsible for another design featuring a type wheel and four row keyboard, but it was only the third effort which proved lucky enough to put in a brief appearance in the market place.

NEAL & EATON

An electric swinging sector design with a four row keyboard was the subject of several US patents granted to these two gentlemen from 1892 onwards.

NEUSSNER

With inventive geniuses of the likes of von Knauss and Jaquet-Droz creating a stir of excitement throughout Europe, a Viennese engineer called Joseph Neussner spoilt the fun in 1783 by publishing an illustrated booklet describing in detail how these magic figures worked.

NODIN

A single example of this appropriately named noiseless typewriter (*no din!*) is in the Henry Ford and Greenfield Village Museum in Dearborn, Michigan.

OSBORN

US patent granted to A. A. Osborn of New York in 1888 for a type wheel design which was not manufactured.

OTTO

Rev. H. J. Otto was granted a US patent in 1907 for a fanciful scheme which entailed piping all the pneumatic typewriters in a building to a single compressor located in the basement. The machines involved were also of his design. Needless to say, the idea was not of divine inspiration.

PAPE

A shorthand machine patented by Henri Pape in 1844 used a system of strokes on paper tape already invented by Dujardin, whose Tachygraphe had put in its appearance six years earlier. Twenty-four keys were used to make their own simple mark on paper tape, whereupon the hinged keyboard was folded back, revealing a scale which corresponded to the order of the letters on the keyboard. The paper tape, previously rewound, was moved past the scale for decipherment. No spac-

133. Cash (Courtesy of This Olde Office, Cathedral City, California)

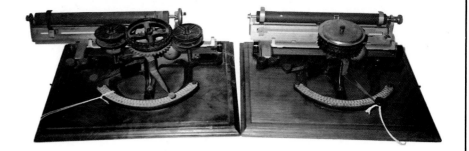

134. Champion (**l**), Peoples (**r**) (Courtesy of Tom Fitzgerald)

ing device was provided; in order to separate the words, the first letter of each word was simply pressed harder, making a thicker or deeper impression on the paper.

PARIS

A French monk called Charles Paris sent as a missionary to China in 1785 was appointed clock maker to the Emperor soon after his arrival and ordered to make a writing automaton such as the one Jaquet-Droz had produced. The product, completed some years later, was a figure between four and five feet tall which wrote texts extoling the virtues of the Emperor in the Chinese, Tartar, Mongol, and Tibetan languages. There is however a suspicion that this was a Jaquet-Droz automaton sent to China from Europe which arrived damaged and was simply repaired by Paris, or that it was adapted by Paris to the Williamson clock now in the Beijing Museum.

PARKER

Roy Parker of New York was responsible for this vertical platen front stroke double keyboard machine for which he was granted a British patent in 1908. In common with other comparable designs by the likes of Hanson and Nickerson, the machine typed around the platen which was raised for line spacing, thereby harnessing gravity to simplify the mechanism.

PARKS & SHEFFIELD

US patent for an electric design was granted in 1883 to these two New Yorkers.

PATRIA (1)

1900, and German typing instructor J. P. Moser who had already designed a pneumatic typewriter with a piano keyboard called Germania was also responsible for a comparable type bar instrument called Patria which used the same keyboard and which dated from the same year and which met the same fate.

PATRIA (2)

Prototypes of a machine by this name were built in Vienna in 1913 by Maschinenfabrik Vulkan but the First World War put an end to the project.

PAYEN

A life-size writing automaton was made by a Frenchman of this name and presented to the Academie des Sciences in 1771, amid great expressions of admiration and acclaim. The figure first lowered his head and then began printing the word *AIMEZ* (a direct ancestor of the modern hippie, perhaps?), raising its hand after each letter and following the writing by turning its head and eyes. A further report two years later notes that the automaton was still at work and had been presented favourably at the French Court. It is known to have used a movement consisting of cams similar in concept to the other writing automata of its day, but neither the instrument itself nor details of the movement have survived.

PEACOCK

A novel British patent for a circular index machine with radial type plungers around a drum and roller inking was granted to E. E. Peacock of London in 1884.

PEELER

An improbable invention, which nevertheless spawned a number of later developments, was the Machine for Writing and Printing patented by Abner Peeler in 1866. A total of eighty-one letters and short syllables were arranged in a 9 x 9 grid. An indicator was suspended above it by cords in such a way that moving the indicator front to back or side to side to select the required letter pulled the cords, which in turn located the desired character on a plate by means of racks attached to its two sides. It was a clumsy design but may well have provided the beginnings from which a number of later index machines

135. Chicago

136. Columbia Bar Lock (Courtesy of This Olde Ofice, Cathdedral City, California)

were designed. Furthermore, by including digraphs on the index Peeler may justifiably claim to be one of the pioneers of the later syllable typewriters.

PELLATON
This eccentric design was invented as late in the day as 1932 by a Swiss watchmaker called Georges Pellaton (a surname much to be conjured with, in horological circles). The electric type wheel invention used a cord wrapped around the type wheel spindle and attached to a pointer which selected letters on an index. Never manufactured, it was doomed even before it left the drawing board as were similarly primitive, albeit non-electric, machines of comparable design patented as far back as 1883 by Perce, the following year by Gilman and Kempster, and the New England and Index Visible manufactured in 1900 and 1901 respectively.

PEMBER
A syllable machine capable of typing up to four letters at a time was patented as early as 1873 by Jay R. Pember of Vermont. Four sets of keys: one set each for the thumbs of each hand, and one each for the fingers. The impression was made by means of a pedal.

PERCE
Patented by W. R. Perce of Rhode Island in 1883, this was the first of several similar designs launched over the years in which a type wheel was turned by means of a cord attached to an indicator which selected letters from a linear scale.

PERROT
A shorthand machine with a separate type wheel for each hand was the subject of a number of patents granted from 1839 onwards to Louis Jerome Perrot. A hammer striking the type wheel spindle brought it into contact with the paper and both type wheels could be used simultaneously. Inking was by carbon paper, which was wound around the platen facing outwards, and the use of very thin or transparent paper so that the impression was visible through it. In days when carbon paper was not the commodity it is now, the alternative of blackening the platen itself was suggested.

PERRY
Charles H. Perry of New York was granted an interesting patent in 1888 for a machine using an index of lenticular section comparable to the Lambert but manipulated by means of a vertical joy-stick.

PETERS
Early comparable inventions frequently have an uncanny habit of bunching together not only chronologically, as evidenced by Unger and Creed, Francis and Ravizza for example, but also geographically. An example of this is provided by the type plunger instrument invented by I. A. Peters of Copenhagen during, or around, the very time that Malling Hansen was inventing his Writing Ball in the same city. Controversy over the credit for this design has continued over the years, fuelled by the uncertainty over the exact dates when the instruments were actually being developed, as against the patent dates which are definitive but considerably later. It has always been my personal conviction that this bunching together of inventions is not attributable to chance alone. I do not by any means deny that inventions may well have a tendency to be 'in the air' at a given time, but technology is more likely to be responsible than telepathy.

Peters filed a patent for his design in 1870 and this ought to be conclusive enough, for by this date Malling Hansen was far enough ahead of him to be actually manufacturing his machine. But Peters was later reported to have been working on his device as far back as 1862 and if that was the case, then it becomes a close thing since Malling Hansen is recorded as having finished his first machine by 1865. In designing, developing, and building a machine of this kind in one's spare time, three years is no time at all.

Whatever the truth behind it all is, the fact remains that Peters' machine used thirty-five slightly curved type plungers located by three parallel horizontal plates to type down on a flat paper table, and the crude large model of the instrument is in the Danish Technical Museum in Helsingor.

PETIT
An 1885 patent granted to N. F. Petit of South Carolina for a design using a circular index with rubber type.

PHILADELPHIA
Typewriter pioneer Byron A. Brooks, better remembered for the typewriter which bore his name as well as his contribution to the development of the Sholes and Glidden in the Remington factory (he invented the shift principle selecting upper and lower case by means of lateral movement of the platen), was responsible for this 1891 type wheel patent with three row keyboard which was to have been manufactured by the Philadelphia Typewriter Co. but never made it. However, the inventor was marginally more successful with another type wheel design using an indicator, which was briefly manufactured as the Crown, and of course with his down stroke type bar machine called Brooks.

PHIPPS
An 1880 patent for an up stroke design was granted to J. H. Phipps of Michigan. It was not manufactured.

PHONAUTOGRAPH

One of several abortive attempts at harnessing the human voice to produce a print-out which a trained operator could read, this one using the vibrations of a phonograph membrane was invented in 1894 by a Californian with the improbable name of A. C. Rumble.

PHONOGRAPH TYPEWRITER

Yet another attempt at a voice controlled machine was invented by Dr. Frank Traver of Racine, Wisconsin. His ambitious project involved type bar selection by means of speaking into a phonograph horn—predictably, at that early date, it was doomed to failure.

PHONOTYP

This shorthand machine sporting a large keyboard of ninety-six keys and typing on paper tape was invented in England in 1924 by E. Howard.

PHONOTYPE REPORTER

J. C. Zachos was granted a US patent in 1876 for a shorthand machine which was to have been marketed under this name.

PIANOGRAPH

A 1902 down stroke musical machine using a piano keyboard designed by a German called Hermann Wasem who learned his craft in the Adler workshops and spent his life in the typewriter industry in one capacity or another.

PORTEFEUILLE

A German journalist living in the United States is said to have invented a typewriter the size of a wallet called, appropriately, Portefeuille. No date or further details are given in the single source of this information, except that the man's name was Anthony Daul.[12]

PRATT

Some interesting typewriters were made by John Pratt of Alabama from 1863 onwards, with British patents dating from the following year. The first model used a type wheel, with carbon paper for inking and a piano keyboard of twenty-seven keys. Noteworthy is the fact that the impression was made by means of a hammer striking the paper against the type wheel, a concept later employed by several mass produced machines, most notably by the famous Hammond. Despite this it was a relatively crude device, as the inventor himself later admitted. The modified second model completed two years later was only marginally better, even though an instrument maker was hired to produce a more satisfactory piece of machinery. The type wheel principle was abandoned on this model in favour

137. Columbia 'double wheel' Type Writer (Courtesy of Sotheby's, London)

138. Commercial Visible 6 (Courtesy of Phillips, London)

of type plungers arranged in conical fashion typing upwards against the paper held in a flat carriage on the underside of the lid which had to be closed for typing. The piano keyboard was retained, as was the carbon paper.

Disgusted with the results of his first two efforts, Pratt now undertook some radical changes. Type bars were out—as so many other inventors both before and after were to find to their dismay, type bars had a terrible tendency to jam at the printing point—and type wheels were not much better because clearly, some letters would have to travel much further than others as the type wheel turned to select the required letter, thus making typing speed irregular. Piano keyboards were out as well, because they were inefficient and required too much effort.

Running the gamut of possibilities (type bars and type wheels already tried and found wanting) Pratt next tried an index design, with thirty-six characters in a square 6 x 6 configuration. Unwittingly blazing yet another trail for future inventors, Pratt reduced the keyboard to six keys per hand, with each hand pressing a key simultaneously to locate the desired character on the plate. A hammer struck paper against type, carbon paper retained. This was the basic specification which he patented in 1866, with an alternative conventional keyboard as an option.

A fourth model was still to come, with sixty-four characters controlled once again by the geometric keyboard, this time with sixteen keys managing an 8 x 8 square index for upper and lower case. After a number of modifications the design reached maturity as Pratt's Pterotype of which a beautiful example is in the London Science Museum collection. Back again to the type wheel principle, with a small diameter wheel and reduced dead-weight allowing for faster operation, it used thirty-six keys for printing the letters of the alphabet and numbers; the keyboard, in alphabetical order once again, was divided down the middle for left and right hands; the paper was still kept flat using carbon for impressions. Despite these anachronisms the model sported some novel features, most notably the use of a spring, which was wound by depressing a knob on the keyboard at the end of each line for the typing action, and spacing by means of the partial depression of any key, plus a suggestion that interchangeable type wheels could be used for different languages and type face.

Pratt's Winged Type, despite all this, did not take off. Several machines were made, but despite lavish praises, no commercial backer could be found to promote it. Some years later a man called Hammond took an interest, however, and spent years completely redesigning it, eventually turning it into the famous machine which bore his name.

PRENDERGAST

J. S. Harrison and H. Hill of Georgia patented an electrically assisted swinging sector design in 1903 which was assigned to the Prendergast Electric Typewriter Co. of Maine but the instrument was not manufactured.

PROCTOR

This was a conventional upright designed by A. F. Mulhare of Baltimore, Maryland, due to be marketed in the early 1920s by a manufacturer of the same city, but the project did not materialize.

PROGIN

'Il y a environ un an que j'ai inventé cette machine' was how Xavier Progin concluded his 1833 patent for his 'machine ou plume ktypographique' with which 'with a little practice,' one could write just as fast as with an ordinary pen. Which was possibly quite true, and the inventor can be credited with a great deal more than mere typing speed. For here we have an early type bar machine of simple and direct design, with the bars arranged in a circle around the printing point, pivoted at the base and rising conically upwards, and thrust down onto the paper by levers attached directly to them. Letters were selected by direct reference to an enclosed circular index of upper and lower case characters, accents and punctuation marks, and by interchanging the index and levers it was even adaptable to typing music. It was the first invention to offer visible typing. Inking was by means of a pad, spacing by racks on two sides, with the paper placed flat beneath the machine. By providing double spaces for his capital letters, Progin also offered a limited form of differential spacing. It was a fine, simple machine and worse designs were still being dreamed up sixty years later.

PROMPTOGRAFO

One of the few Spanish contributions to the development of the typewriter was an 1881 patent granted to Vincente Alonso de Celada y Barona of Barcelona for a stenographic machine of fifteen keys printing on paper tape at a speed claimed to be 140 words per minute.

PROUTY & HYNES

Full credit is rarely given to Enoch Prouty and Olive Hynes for their US patent filed 5th January, 1887 (British patent dated 15th November 1886). It was the first front stroke type bar example of what was destined to become the conventional upright. Four row keyboard, with ribbon (but the ribbon spools were located in the middle of the machine and the ribbon travelled not from side to side but front to back), the type bars located horizontally in a segment in front of the platen.

RAAB

W. Raab of Iowa was granted a US patent in 1894 for a pneumatic up stroke machine using a keyboard of rubber balls, three years after Weir patented his virtually identical Pneumatic.

RAVIZZA

There were many inventors whose involvement with the development of writing machines was peripheral or even tangential. Others however, became involved on a deeper and more intense level, while a few may be said to have dedicated the better part of their lives to the machine. Some of these men were ultimately rewarded. Others, tragically, never saw their dreams materialize, nor their efforts crowned with success or even recognition.

One such man was Giuseppe Ravizza, a true pioneer of the writing machine who was to spend fifty years of his life unsuccessfully grappling with the problem. And as if this were not punishment enough, he was destined to live long enough to see the typewriter establish itself in the United States with features he himself had patented thirty years earlier, (even though, if the truth be told, he did not himself add a single original component to typewriter history). An entry in his diary, some months before his death in 1885, said it all: 'By now I am beginning to despair of this poor machine, bane of my life. Although I am so close to success, seeing that my health shows no sign of improvement I fear I have not long enough to live.' By then, Ravizza had built sixteen models and although details of all of them have not survived, we do have enough information to know that they were all different, and all in one way or other embodied modifications which he hoped would solve his problems.

His 1855 patent, itself a product of eight years work, was the first tangible confirmation of his efforts, inspired perhaps by his earlier correspondence with Conti, the inventor of the Tachigrafo. The Cembalo Scrivano, as Ravizza called his machine, was of blind up stroke design with type bars in a circular basket beneath the flat paper table. (That feature itself was far from original to Ravizza, nor was he the last to use it—as we have seen, Francis, Mitterhofer, Sholes, and many others designed instruments similar in concept.) It had a piano keyboard of thirty-two keys, paper table advanced by clockwork, inking by means of a silk sheet impregnated with rust or graphite. Refinements well in advance of his time included conical type bar pivots to prevent jamming and ensure alignment. A cylindrical platen was suggested as an alternative, printing around rather than along its length.

140. Competitor (Courtesy of Bill Kortsch)

Ravizza's machines were displayed quite extensively in Italy. They were exhibited at the principal Exhibitions in at least four cities, loaned to friends, colleagues, and possible promoters, and a few were sold. But commercial success eluded him and not even important developments such as a model with visible typing which he had produced as early as 1872, nor an open model which reduced noise (1879), nor even his sixteenth and final model (1882), with which he expected to reclaim the growing local typewriter market from Remington, were able to produce a change in his fortunes.

It was the Remington, 'based on my own principles,' succeeding where he had failed, which proved his ultimate heart-break. **(15)**

REED

Charles J. Reed was granted a US patent in 1892 for an electric machine which was not manufactured.

REICHSFELD-PAD

Alfred Reichsfeld of Vienna is reported to have been responsible for a machine by this name which he exhibited in the Austrian capital in 1899 but no description of the device appears to have survived.

REMMERZ

There is a sole reference in a generally unreliable source to an indicator machine invented by the German brothers H. & F. Remmerz which used a separate type index for upper and lower case and figures.[12] Efforts at tracking down this invention have been unsuccessful.

RICHARDSON

H. B. Richardson of Massachusetts patented an indicator machine in 1884 which sported a type wheel of elliptical profile mounted on a horizontal arbor.

RIOTOR

Little is known of this machine other than that it was French, dating from 1908, that it had type bars resting on a vertical inking pad, and that it typed by bringing the type bars down onto a horizontal plane. The description may be of some form of grasshopper action, or, alternatively and possibly more likely, of a design in which the type bars twisted through 180° on

139. Commercial Visible 6 (Courtesy of Bernard Williams)

141. Corona 3 (Courtesy of This Olde Office, Cathedral City, California)

their way from pad to platen, in which case the correct name may well have been 'Rotor.'[81]

ROBERTS

An 1884 British patent for a shorthand machine with twenty keys. A further patent the following year cut the number of keys to nine, of which two or more could be used simultaneously.

ROE

1885 type wheel patent granted to E. R. Roe of Illinois.

SAINTARD & SAINT-GILLES

In 1845 two Frenchmen, Dr. Saintard and his mechanic Saint-Gilles, patented a frame to assist the blind to write. It was later developed through additions to the original patent into a circular index writing machine printing twenty-eight characters.

SALIGER

Alois Saliger of New York patented a stenographic type bar machine with a keyboard of twelve keys in 1904.

SAMENHOF

A single reference to a machine said to have been invented by one Dr. L. Samenhof using an indicator and an index in the form of a comb with the letters on the flexible teeth.[12] Further details are lacking.

SANGER

A shorthand machine, subject of a British patent dated 1871.

SCHAPIRO

An indicator attached to a rack selected letters from a linear index, the lateral movement of the rack turning a type wheel with which it was meshed, and a separate key brought type wheel down onto the platen below. A. Schapiro of Berlin was responsible for this 1894 invention.

SCHIESARI

A syllable typewriter featuring the ability to print up to six letters at a time by means of the simultaneous depression of six keys—this 1905 invention by the Italian Mario Schiesari was to have been manufactured in 1914 by The Syllable Typewriter Company of New York but the project did not materialize, and in this case, the outbreak of war may not have been the only reason.

Schiesari's complex project specified type bars located vertically behind the platen, printing by down stroke, the type bars to be displaced laterally by up to 12.5 mm from the centre if a maximum of six keys was pressed simultaneously, the carriage automatically advancing the correct number of spaces. Like other similar syllable machines, in which keys were played like chords on a piano, it suffered from the obvious complication that the typist had first to work out whether the letter order in the syllable to be printed corresponded to the letter order of the keys on the typewriter, so that in the end it was infinitely slower than the conventional machine it was designed to supersede, however long and hard the typist practised.

SCHMITZ

A shorthand machine with five keys printing by up stroke, the shorthand code consisted of cells composed of up to three horizontal and two vertical strokes. Patented in Germany in 1882.

SCHOCH

Type bars fitted with pens which printed a shorthand code of dots on paper tape advanced by treadle or clockwork was the contribution of an 1870 patent granted to Henry Schoch.

SCHÖNFELD

A musical writing machine was reported to have been built in 1808 by a Pastor Schönfeld. Like others at the time, it was designed to be fitted to a piano keyboard.

SHEEHY

In 1884 a patent was granted to R. J. Sheehy of New York for a machine in which a type wheel, controlled by a pin barrel, moved along the paper which remained stationary.

SHEFFIELD

The Rev. D. L. Sheffield, an American missionary in China, made an early attempt at producing a typewriter capable of typing 4,000 of the most common Chinese characters. The 1899 design called for a large horizontal circular index with the characters on the upper side corresponding to type beneath. Paper was in a carriage under this wheel which was turned to the desired line and the carriage positioned so that the correct space corresponded to the character selected, with a hammer forcing the paper against the type face.

SHERMAN

This interesting down stroke machine patented in 1877 by Daniel H. Sherman of Ohio included features heralding later book typewriters. Semi-circular four row keyboard, printing by ribbon.

142. Corona 'Special' (Courtesy of Christie's South Kensington, London)

SHERWOOD

Edward Sherwood of Rotterdam invented a blind type wheel machine with automatic carriage return and keys which stayed down until each typing operation was completed but his 1901 application for a patent to protect the invention was unsuccessful.

SHINOWARA

A Japanese government official by this name is alleged to have invented a machine in 1908 which was approximately 30 cm square, and capable of typing 2,500 different Chinese or Japanese ideographs.

SHOLES CHRISTOPHER LATHAM

An experimental model of a machine designed to replace troublesome type bars with a type wheel or type sector was built by Christopher Latham Sholes, whose involvement with typewriters continued long after Remington took control of his Sholes and Glidden Type Writer. The model is preserved in the Milwaukee Public Museum and illustrates Sholes' conviction that type bars would never produce perfectly aligned work.

Sholes' 1889 patent protected a design featuring an octagonal type sleeve, which failed to reach manufacture, and in fact only the unconventional Sholes Visible, of all his designs apart from the first, actually made it past the drawing board.

SHOLES F.

Fred, one of the sons of Type Writer pioneer Christopher Latham Sholes, was granted a patent in 1880 for a down stroke machine in which the type bars were located behind the platen.

SHOLES F. & MILLER W.

An 1879 patent for an up stroke in which the horizontal type bars fanned outwards from the printing point was granted to Fred Sholes and William Miller of New York. The principle feature of the design was the manner in which the front of the type bar swung up against the platen as the rear of the bar dropped down a slope, tension on a spring returning it to its original position.

SHOLES L.

Louis Sholes, another of Christopher Latham's sons, is credited with a conventional front stroke machine with four row keyboard, the prototype of which is in Milwaukee.

143. Crandall (Fennia) (Courtesy of Sotheby's London)

144. Crown (Courtesy of Bernard Williams)

SHOLES Z.

The most successful of Christopher Latham's sons was Zalmon who is credited with a front stroke machine with three row keyboard, of which a prototype survives, but whose major contribution was his involvement with the successful Rem-Sho **(225)**.

SIEMENS

A contribution to the field of printing telegraphy was made in 1850 when Werner Siemens designed a machine using a type wheel of radial segments similar to the one patented by Wheatstone some years earlier. It was just one of the inventions of this pioneer of telegraphy who, along with his mechanic J. G. Halske, went on to build up one of the most important electrical organisations in the world. Later printing telegraphs used a perforated tape system.

SILKMAN

E. J. Silkman was granted a US patent for an electric machine in 1890.

SLOCUM

W. H. Slocum of New York deserved a better fate than he received for several of his interesting designs which were never manufactured. A type wheel machine printing with the assistance of a hammer comparable to the Hammond was patented in 1879. The two row keyboard printed upper and lower case and offered the additional sophistication of differential spacing.

In 1886 he patented an up stroke type bar machine comparable to one patented ten years later by Marshman and Burridge and manufactured as the American. Slocum's design was better, however, in that the type bars were pivoted and guided to the platen.

SMITH J.

A single reference to this machine, invented by Joel Smith in 1865, is reported in the proceedings of the 1963 International Congress on Technology and Blindness.[47] Called a Daisy Point Writer because its six keys were in the form of flower petals, the machine was designed as a small block which rode on rails, pricking the page held on a flat base.

SMITH R. H.

Robert Henry Smith was granted a British patent in 1900 for a stenographic machine printing a phonetic alphabet based on only two characters used separately or together in different positions.

SOBLIK

A remarkable invention using compressed air supplied by two foot pumps was patented in the United States in 1899 by Max Soblik, a German engineer living in Belgium. The keyboard consisted of rows of holes through which the compressed air passed. The typist simply placed a finger over the desired hole

145. Cub Reporter (Courtesy of Sandy Sellers)

and the air supply to that hole was diverted from the keyboard to the corresponding type plunger located in a wheel. Operating speeds were claimed to be four times faster than conventional machines. Dispensing with the hardware of the usual typewriter apparently made for silent operation, although if ordinary turn-of-the-century foot pumps are anything to go by, the noisy keys may well have been replaced by the wheezing and creaking of bellows.

SOENNECKEN

A machine which appears to be similar to a Merritt in operation was attributed to F. Soennecken of Bonn.[12] It appears to be either a linear index or a linear type plunger machine, and was claimed to be similar to a Schapiro, but this would appear to be manifestly incorrect. No date was given and efforts to trace this invention have been unsuccessful.

SPIRO

Charles Spiro, a New York watchmaker, was a prolific typewriter inventor with famous machines such as the Columbia and the Bar Lock to his credit, plus lesser designs such as the Visigraph and the Gourland. Not all his efforts were successful, however, and those that were patented but not produced include an 1885 linear index machine using a keyboard, plus numerous type wheel mutants dating from 1886.

STEINHEIL

A printing telegraph invented by a German professor in 1838 used two arms fitted with inking devices to print a two row code of dots. This ultimately lost out to Morse's rival system because of the latter's easier use with audible signals.

STENO

A shorthand typewriter using two pairs of type wheels was the subject of a 1922 patent which was not manufactured. One pair covered opening vowels and syllables and the other was for closing vowels and syllables.

STENOGLYPHE

A French lieutenant by the name of Muller invented this shorthand braille machine in 1900 or so.

STENOTIPOSILLABICA

A trained Michela stenographer attached to the Italian Senate was responsible for this piano keyboard machine clearly inspired by the Michela invention—a piano keyboard in two parts separated by the paper tape but using a total of forty-six keys. Each key operated its own type wheel around which the character was repeated seven or eight times, with a ratchet turning the wheel after each impression. The inventor was Dario Mazzei and the year was 1879.

146. Dactyle (Courtesy of Musée National des Techniques C.N.A.M., Paris)

147. Darling (Courtesy of Sotheby's, London)

STENO-TYPOGRAPH

A German Remington representative called Angelo Beyerlen was responsible for an 1884 telegraphic shorthand machine which featured a piano type keyboard of five keys per hand.

STERLING

An inexpensive version of the Bar-Lock typewriter to be called Sterling was announced by its manufacturer, Columbia Typewriter Co. of New York, but the scheme appears not to have materialized, possibly because the name was already protected by the unrelated manufactured machine.

STICKNEY

Burnham Stickney of New Jersey was granted an 1896 patent for an up stroke machine which spaced just before the type face struck the platen.

STÜBER

Dr Werner Stüber of Leipzig designed a syllable machine with sixty-six keys printing on paper tape, but three prototypes, dating from 1936 on, were destroyed during the war.

TACHIGRAFICA

This shorthand machine based on the Stenotiposillabica of Dario Mazzei was the work of a man called Bussadori from Bologna. The year was 1880.

TANGIBLE

This appropriately named typewriter for the blind producing embossed braille cells was due to be produced in England in 1895 by the Tangible Typewriter Co., but the machine was never manufactured.

TARABOUT

A complex design of a shorthand machine featuring a grasshopper action was patented in France by Aristide Tarabout in 1904—paper tape and inking by pad.

TAYLOR

Joseph Taylor of Rochester invented an electric machine in 1910 for which extravagant claims were made but the machine was not manufactured.

TELEELECTROGRAFO

Luigi Lamonica's shorthand machine, reportedly perfected by Innocenzo Golfarelli of Florence in several instruments made from 1875 for use in telegraphy. Keys, pressed singly or in combinations, were used to print a shorthand code of the inventor's own design on paper tape which was at first fed through the machine 'vertically' (as on modern shorthand machines), but later horizontally, permitting the message to be read like conventional typing.

TELESCRIPTOR

This printing telegraph device used a single row of characters on a type wheel of ample diameter printing on paper tape. It had a conventional keyboard, with an electrical contact under each key. Clockwork supplied the motive force. An 1896 device invented by M. Hoffman, it was said to be capable of handling 120 characters per minute.

THOMAS H.

H. J. Thomas of New York patented a transverse linear index machine in 1886. The index was of square section and had type face on all sides, permitting syllables and short words to be printed at one stroke; this hybrid of a syllable machine and a linear index was quite remarkable, but apart from this feature, the transverse linear index design was manufactured with some success in machines such as the Sun **(255)** and the Odell **(202)**.

THOMAS R.

A Typograph patented by Robert Thomas in 1854 consisted merely of type attached to a cylinder and moved manually in a wooden frame, and discounting the Writing Glove, was quite possibly one of the most rudimentary typing devices ever invented **(42)**.

THURBER

The much maligned Charles Thurber of Worcester, Massachusetts, made a number of valuable contributions to typewriter history, the details of which have caused considerable confusion in books on the subject. In fact, when I began researching some twenty years ago, I was amazed to find that every earlier source was incorrect. However, despite my determination to set the record straight in *The Writing Machine*, I now find that I also slipped up on one of the details.[1]

Thurber's first contribution was an 1843 patent for 'a new and useful Machine for Printing by Hand by Pressing Upon Keys Which Contain the Type,' called Thurber's Patent Printer. It was useful for the blind and the nervous 'who cannot execute with a pen.' The design called for type plungers to be fitted around the perimeter of a horizontal wheel mounted on a vertical spindle—as many plungers as were necessary, although his model contained only fifteen 'inasmuch as it was designed merely to show the principle.' Inking was by roller—'when a line is ended, whirl the wheel around to get fresh ink'—and printing was on a flat carriage. This first machine already incorporated differential spacing whereas I had previously attributed that feature only to Thurber's third machine. Meanwhile, a second model of the Patent Printer which exists today in the collection of the Worcester Historical Society already boasts a cylindrical platen to replace the flat paper carriage, but printing was performed around the cylinder rather than along its length as on modern machines. Thurber was thus one of the first to use a platen. The machine was incomplete when discovered and some of its components are believed to date from its restoration.

A third Thurber machine which he called his Mechanical Chirographer was an abrupt departure from his earlier inventions. Patented in 1845, the device used a pen or pencil to print by following the motions a hand would use for the same task. This was very much an 18th century concept; it produced quite presentable work (so did they) but according to some 'it did not answer well,' although not everyone appears to have shared this view.[50] A contemporary eye-witness report was printed in the Philadelphia 'Public Ledger' on 28th February, 1846, and is worth reproducing in its entirety:

'An Ingenious Invention—We noticed a day or two ago, the fact that a machine had been invented by Charles Thurber, Esq., of Norwich, Connecticut, by which the blind may be taught to write with the same facility as those who can see. We have since seen the machine, Mr. Thurber being now in this city, at Sanderson's, and it is a very ingenious invention indeed.—The instrument resembles a small piano; each key is marked with a letter. The keys are connected with machinery in the interior of the instrument, by which both a perpendicular and a lateral motion are obtained; these make the pen, connected also with the machinery, form the letters like small Roman capitals. The keys are struck by the fingers precisely as in playing on the piano forte, and the pen, which is supplied with common ink, makes a letter at each touch with the finger, on a sheet of paper fixed up in front of the instrument. There is another key to make the spaces between the words, and likewise one, which being pressed in conjunction with another, makes the letter at the beginning of the word, if desired, larger than the rest. The machine is so constructed that when the paper has traversed its proper limits, it is forced back to its original position, raised sufficiently high to leave the necessary space between the lines without any effort on the part of

the operator. A description of the machine, unless in its minute detail, would scarcely convey to the reader a proper idea of it. In its operation, it succeeds in making every letter almost as fast as it can be done with the pen of a quick writer, and as clear and distinct as printed letters. It not only would suit the blind, but, in the last particular mentioned, would be a desirable instrument for some who can see.'

A fourth machine was built but has not survived. **(40, 41)**

TILDEN-JACKSON

The Tilden-Jackson Typewriter Company of Hamilton, Ontario, with half a million dollars of capital stock was set to manufacture a machine but the project was abandoned in 1908.

TILTON

A linear plunger patent with index and indicator, granted to C. E. Tilton of Massachusetts in 1884 with inking by ribbon.

TIPHLOTYPE

This braille machine using six keys to emboss braille cells was presented in Amsterdam at the 1885 Exhibition for the Blind. A Russian count by the name of Kovako designed it but it was not manufactured.

TIPO-STENOGRAFICA

An old idea which re-appeared more than somewhat, the Tipo-Stenografica invented in 1888 by Luigi Ranieri of Rome used keys pressed simultaneously to build up letters from their component strokes. It was no more successful than any other similar device.

TOEPPER

Richard Toepper, who patented an up stroke syllable machine in Germany in 1896 which was to be called Blitz, was also responsible for an index machine to type syllables which he designed the previous year. Neither was manufactured.

TOLLPUT

The 1851 Exhibition introduced some great machines to the world, not the least of which was the Hughes Typograph **(2)**. An 'apparatus' for the blind invented by a man called Tollput was merely mentioned in the Exhibition Catalogue but no description was given.

TURRI

There are no monuments to this man. No statues, no anniversary celebrations, no memorials...not even a plaque. And yet Pellegrino Turri (1765-1828) was the inventor of the first machine which typed letters of the alphabet—the first man to make such a machine which we know to have worked. The

148. Daugherty Visible (Courtesy of Bernard Williams)

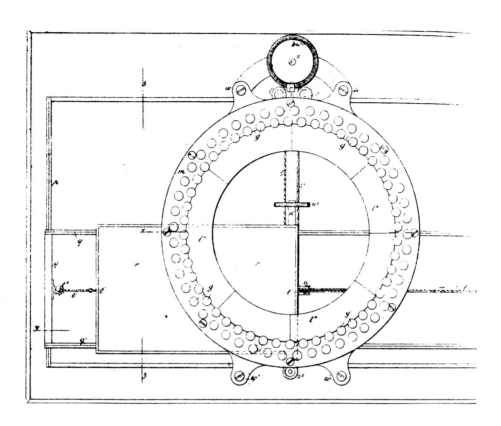

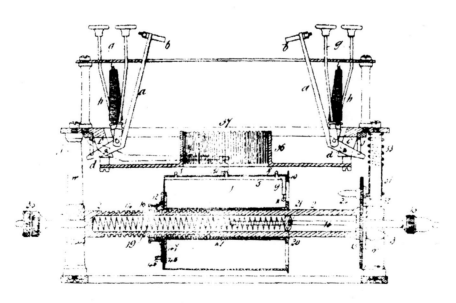

149 (a & b). Daw and Tait (British patent)

Rev. Creed's musical invention was the first known 'Writing Machine,' but Turri's was the first 'Typewriter.'

I stumbled across this man's name a long time ago, very early on in my collecting days. I had no reference books on the subject, and as far as I knew, there were no other collectors in the world, such was the vacuum in which we few typewriter collectors in the world lived in those days. My information came not only from patent dates on the machines I found, but also from old magazines and newspapers which carried advertisements, illustrations, and articles, and the relevant machines could thus be dated. In an article by one Umberto Dallari entitled 'Nihil sub sole novum' in the March 1908 issue of the Italian magazine La Lettura (I was living in Rome at the time), the writer referred to the curious fact that typewritten correspondence exactly 100 years old existed in the State Archives in Reggio Emilia.[56] A photograph of an original typewritten page illustrated the article. I duly noted all the facts, without in any way appreciating their momentous significance, and some years later, when I 'discovered' Johannes Meyer's book, I found a few lines of reference to Turri in his section on typewriters for the blind.[63]

It was only very much later that my research revealed the true significance of the matter. For this was by far the earliest proof of the existence of a type writing machine—and what proof it was! The Reggio Emilia Archives had a file containing page after page of correspondence typed on the machine.

The family's papers were all handed over to the Archivio dello Stato after Turri's son, Giuseppe, died in 1879. Umberto Dallari, the director of this institution in the early years of this century, was sorting through the correspondence when he was struck by the curious fact that the letters he was reading had been typed exactly 100 years before. So he wrote an article about it for La Lettura, entitled 'Old Letters Written by Machine.' He made no claims about the correspondence. It was simply a curiosity.

But although Turri made the machine and it was in his file that the papers were found, they were not written *by* him but *to*

him. Countess Carolina Fantoni, struck down by blindness in the 'flower of her youth and beauty,' was in her early twenties when Turri built the writing machine which permitted her to correspond with him. Why he should have gone to all this trouble when she could so easily have dictated her letters to a secretary might lead one to suspect that they were both more than a little anxious not to publicize their correspondence, a view given even more credence by the fact that the Countess suffered crippling arthritis of the hands and paralyzing headaches, and as she so often complained in her letters, using the machine could not have been easy for her. It is a credit to Turri's invention that she was able to use it at all and, under the circumstances, the results are remarkably good. They are as good, in fact, as any typing on any other machine until the 1870s.

The machine itself has unfortunately not survived, so that the only information we can glean about it is from the correspondence. From this we learn that first of all carbon paper was used, and later, some form of inking system was substituted since Carolina had difficulty obtaining the carbon paper which in those days, of course, had to be made by hand and was not usable after a couple of applications. Some of the carbonned letters are faint indeed. On the other hand, the use of ink proved little more successful. It was messy when first inked, and then progressively fainter as the typing continued: when first inked, it bled through the thin paper often making the typing barely legible, since Carolina would insist on using both sides of the page. But for all that, considering its early date and the disabilities of the user, it must have been an outstandingly good machine. Maybe it has survived, discarded in a loft somewhere in Italy, and may yet see the light of day. We can but hope. **(35, 36)**

TYPEN

Comprising a type wheel turned by a knob and suspended in a wire frame bent into a handle at the top with roller inking, Dr. Henry Wetherill of Audubon, Pennsylvania, patented this machine in 1924.

150. Diamond (**l**), National (**r**) (Courtesy of Bernard Williams)

151. Edison Mimeograph (Courtesy of This Olde Office, Cathedral City, California)

to take a more benign view, that he independently developed an identical instrument which he communicated publicly for the first time, however, only four years later and two years after the publication of Creed's invention. One way or another, it makes him only the second man to have invented a writing machine.

Grove's definitive *Dictionary of Music and Musicians* (5th Ed.) lists Unger and refers to the machine as a Melograph, a description of which was published in 1774 under the title '*Entwurf einer Maschine.*'

UNION

Lee S. Burridge of New York who sired a host of inventions but is best remembered for the Sun and the type bar American, patented a down stroke machine with three row keyboard in 1900 which he assigned to the Union Type Writer Co. of New Jersey.

UNIVERSAL

A patent granted in 1902 to Otto Rossler for a syllable typewriter using type bars and another granted the same year to Carl von Sudthausen for a syllable machine using a swinging sector formed the basis for a syllable and word typewriter which Dr. Franz Joseph Müller and the Universal-Silbenschreibmaschinen GmbH of Berlin was to produce in Berlin in 1904. The project did not proceed past the experimental stage.

U S TYPEWRITER

C. Wing of Greenfield, Massachusetts, was granted a patent in 1887 for an indicator machine using type on the teeth of a linear comb inking by roller and the instrument was to have been manufactured by the C. Wing Machine Works of the same city but the project appears never to have materialized.

VAIL

Several contributions to the history of printing telegraphy were made by Alfred Vail from 1837 on, with designs for machines which printed the alphabet instead of merely a code. His third effort, and the one for which he is best known, was an instrument employing type plungers fitted radially around a vertical wheel and printing on paper advanced by clockwork.

VELOGRAPHE

A French patent for a double keyboard syllable machine comparable to the Duplex on which two keys could be pressed simultaneously was granted to Riom and Carabasse in 1899. It had a total of ninety keys, with opening single and multiple consonants on left, final consonants and vowels on right. Speed of operation was claimed to be sufficient to permit its use for stenography.

VILLEY

In 1916 a blind Frenchman by the name of Pierre Villey invented this machine which used twenty keys to print on paper tape.

VISBECQ

Roger Visbecq's 1929 musical typewriter drew its inspiration from electro-chemical inventions predating it by a century. It used a piano keyboard with an individual electrical contact beneath each key. The contacts were wired to corresponding levers acting on specially-prepared moist paper over a metal roller so that pressing a key closed the corresponding circuit and caused a mark to be made on the paper between the tip of the lever and the metal roller at the point of electrical contact.

TYPO-STENOGRAPHE

Smitter and Legros were granted a French patent in 1877 for a shorthand machine printing vowels on paper tape as a code of dots above and below consonants.

UHLIG

One of typewriter history's most prolific inventors was a resident of New Jersey called Richard W. Uhlig who over the years designed some fifty or so machines of which a dozen or more were manufactured. Among these were such well known makes as Emerson and Commercial Visible. Among the unknowns is one to which he was to give his name, which was to have been manufactured by the Uhlig-Gunz Co. in 1910.

UNGER

Johann Friedrich Unger (1716-1781), a German lawyer and one-time mayor of Einbeck and councillor of justice in Brunswick (a respected and prominent gentleman, it would appear), claimed the invention of a musical writing machine to be attached to a keyboard instrument in such a way that it recorded all notes played as a series of longer and shorter strokes on paper passing beneath the keyboard.

There is no doubt about Unger himself, nor the instrument he claimed to have invented, but controversy surrounds the dating of the device. Unger first made his invention public in 1749 and again in 1752, claiming, however, that it harked back to 1745. This date would have made it two years earlier than an identical instrument devised by Reverend Creed whose invention was first published posthumously in 1747; he was therefore in no position to dispute Unger's claim.

But why, one is tempted to ask, did Unger delay for four years from the alleged date of his invention before publicizing it, two years after Creed's death...which, one is tempted to suspect, might well have been the time it took for the news of the Englishman's demise to reach him?

It may well have cost him an important historical advantage, for if his claim were indeed genuine, Unger would now be honoured as the first man to invent a writing machine. However, typewriter history is replete with litigants making similar unsubstantiated allegations and in view of the fact that Unger's belated pre-dating of his invention is not graced by any form of corroborating evidence, one is obliged to take the view that either he plagiarised Creed's invention or else, if one wishes

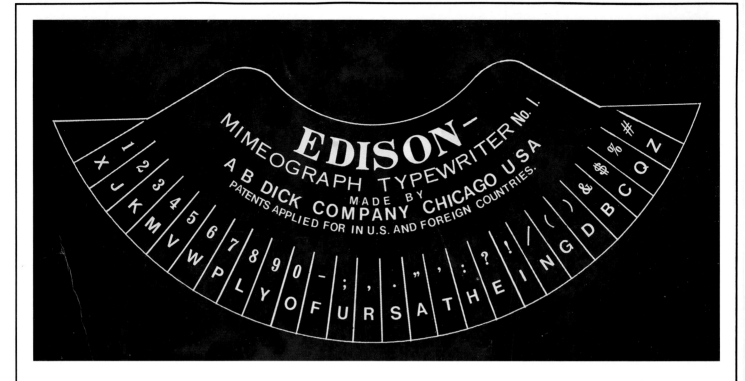

152. Edison Mimeograph No. 1 Index

VOYANTE

A five row keyboard controlling a corresponding five row type wheel was at the heart of this 1898 French design which never left the drawing board.

WAGNER

The prolific Franz Xaver Wagner, best remembered for the invention of the machine which was to become the Underwood **(75)**, participated in the design of numerous other instruments some of which he patented himself and others which he assigned to his associates. Yost **(60, 278)** and Densmore **(95)** are two of the manufacturers to whom he contributed his mechanical skills.

His patents included one in 1880 for a swinging sector machine with four row keyboard, another four years later in which the four row keyboard controlled a swinging segmental comb, and yet another, in 1888, for a type wheel with linear index.

WAITT

G. L. Waitt patented a novel design in 1887 in which type was fitted to the underside of flexible arms which radiated outwards like spokes in a wheel. The patent was assigned to the Edison Type Writer Co. of Philadelphia but the instrument was not manufactured.

WALKER

An ex-employee of the Union Typewriter Company by the name of C. Wellington Walker of Stamford, Connecticut, was responsible for this invention, leaving his job in 1909 after eighteen years with the company to promote it. The machine he invented the following year was said to be noiseless, ribbonless, and basketless in addition to having a motionless carriage. It was also productionless—the inventor sold out to a Boston concern in 1912.

WALLACE

Three row portable with double shift, the subject of five years of experiments by a Columbia University chemistry professor by this name from 1912 onwards.

WARD

Caleb Ward of New York patented this two row circular keyboard up stroke design in 1878 and 1879. Inking was by roller and the platen was positioned in the centre of the machine.

WEBSTER

A tiny portable housed in a leather case measuring a mere 8 x 6 x 2 inches, with a four row keyboard to which were added a few special keys each typing short phrases such as *'Yours truly.'* Joseph March Webster of Liverpool was responsible for this fanciful design in 1898.

WESTPHALIA

Ernst Wilhelm Brackelsberg was responsible for this linear type plunger machine which he patented in 1884 and named after the province in which he lived. A handle on the side of the machine served for letter selection from a linear index. Carbon paper for the impression. The inventor was later responsible for a syllable design capable of typing up to four letters simultaneously (see Brackelsberg).

WHEATSTONE

Sir Charles Wheatstone, who was responsible for so many contributions to the fields of science and technology can lay claim, wholly or partially, to no less than six separate designs of writing machines and printing telegraphs, spanning some two decades and dating back to a single 1841 patent.

Teeth cut around a spring steel disk bore the characters and were brought into contact with the paper by means of a tap from a hammer controlled electro-magnetically. The design called for a cylindrical platen; a flat paper table controlled by rack and pinion was suggested as an alternative. The cylinder travelling along a screw meant that, in common with many another design of the time, printing was performed around, rather than along the page. Carbon paper was specified.

Modifications to the design were introduced in successive models, with mixed success. Type on a square metal comb

153 (a & b). Edland

was tried, and in 1851 a piano keyboard was added and a type sector substituted for the square format. Inking was now by means of a pad and printing was on paper tape. It was not until 1856, however, that we find a radical departure in design: the keyboard consisting of cylindrical keys was now positioned directly above the selector mechanism giving the machine a vertical rather than horizontal aspect. Upper and lower case made the instrument more complete, but it still typed around the platen and both the comb and previous inking systems were retained.

It was Wheatstone's penultimate effort. He tried once more with a model combining different features of his previous designs but with no greater success. Details of Wheatstone's relationship with Bain have already been mentioned and there is every reason to believe that Wheatstone was in fact guilty of plagiarizing Bain's initial work in this field. How much he borrowed from others in the intermediate machines is speculative, although he is known to have worked with a different partner on the 1856 model, which may help to explain its radical departure in design.

WHYTE

A machine which typed around a vertical platen riding up a spiral groove was the subject of a British patent granted to J. G. Whyte in 1903.

WIEDMER

Conventional portable of tiny proportions first exhibited in 1907 and due for production the following year by Deutsche Schreibmaschinenfabrik H. Wiedmer & Co. of Bruchsal. Only a few prototypes were made.

A musical typewriter was also announced by the same company in 1908 but did not reach manufacture.

WIGHT

A machine for the blind using nine keys riding on rails to produce embossed dots on paper held in a flat frame, was built by an American called Joel Wight in 1850.

WILKINS

An 1885 US patent granted to Timothy Wilkins of Iowa protected a book typewriter in which the machine, comparable to the Hall, rode across and down the page on racks.

WIRT

This wondrous machine with its large sixteen sided type sleeve capable of printing up to 280 different letters and musical characters was designed by an American called Wirt in 1904. A combination of shift keys and a twenty-eight key keyboard controlled the operation of the type sleeve, with a hammer striking the paper against it. It used a ribbon. It was said to have been manufactured but no examples have survived and no confirmation of manufacture has been uncovered.

WORLD FLASH

A battery operated electric similar to the Remington was said to have been patented by The World Flash Co. of Chicago but the device was never manufactured.

WORRALL

T. D. Worrall of Washington patented a machine in 1886 with a five row keyboard, consisting of pivoted levers with keys at one end and type at the other. It had a flat paper table, with the keyboard and type bar assembly propelled above it by coil springs. The patent was assigned to the Worrall Manufacturing Company of Boston but does not appear to have been produced. Apocryphal references to an instrument by a man called Worral *(sic)* may well be dealing with the same device. **(109)**

WRIGHT SPEEDWRITER

A conventional front stroke machine made of aluminum and bearing this name is extant, having come from the Salvation Army in San Francisco. No other details are available and no other example is known.

YEREMIAS

Two electric machines, the first with a conventional and the second with a double keyboard, of simple design requiring key depression of only 1mm, were invented in 1899 by a Hungarian called Arnold Yeremias but not manufactured.

YOUNG

A vertical type wheel geared to a horizontal indicator was the design which J. L. Young of New York patented in 1883.

ZEROGRAPH

This was a printing telegraph machine patented from 1895 by Leo Kamm of London which was hailed as the 'most wonderful' of all similar instruments and was received with great enthusiasm in many countries but was never manufactured.[62] The only known example to have survived was previously in the author's collection—a beautiful machine in brass and mahogany with a curved two row keyboard and a swinging segmental comb of Wheatstone inspiration. The machine was of start-stop operation with the added feature that the type sector swung back to the extreme left after each letter was printed and transmitted, thereby zeroing itself out and giving rise to its name. This zeroing effect was performed with electromagnetic assistance, while a spring was used to swing the sector around in preparation for printing. Pins operated from the keyboard limited the swing of the sector for letter selection and a hammer striking type against paper printed the corresponding letter. It was the first machine to apply the start-stop principle to printing telegraphy.

Chapter Six—
Directory of Manufactured
Typewriters

AUTHOR'S NOTE: An attempt has been made in the case of almost all of the following entries to provide a price guide to the current market values of the machines in question. This is a most hazardous undertaking, fraught with danger in every entry. For so many reasons—some blatantly obvious and others more obtuse—the prices quoted must be used only with the greatest circumspection and in the broadest possible sense, in order to establish the approximate parameters within which the monetary value of a machine might reasonably be expected to fall.

Collectors themselves are often notoriously unreliable as a source of information on market values, as has already been noted in the Introduction, and their fragile egos are not always the reason. Some swear blind that they pay next to nothing at all for their machines in the hope that they will be able to buy more cheaply as a result. Others are the exact opposite and boast that they are the world's most extravagant spenders, enhancing the prices they claim to have paid in the hope of being given first refusal on all future offers.

Auction prices can be important tools for determining values but must also be used with extreme caution because they are likely to be dependant not only upon the reserve price placed on the machine by the vendor, which itself might be unrepresentative, but by the artificial *macho* factor which can so easily inflate prices paid when two affluent hydrocephalic collectors, in a cost-no-object exercise, lock horns and vie with each other for the prize they both covet. The resultant maverick hammer price does not determine the true market value of that make and model of machine, even though it can temporarily throw accepted knowledge and wisdom into utter confusion, compounded at a subsequent auction when the unopposed former under-bidder, now without the competition, buys an identical machine for a fraction of the previous amount. To quote but one example: one of the world's rarest and most desirable typewriters is a Hansen Skrivekugle. The hammer came down (metaphorically, of course) on one of these at an auction in 1989 for approximately $40,000 (+10% auction commission) while another remained unsold at less than $25,000 the following year. Of course, on the other hand, all of us recall cherished moments when we bought an article at auction for next to nothing because we had a little more knowledge than did both those who attended the sale as well as the auctioneer who had inadvertently catalogued the lot incorrectly.

Furthermore, despite claims to the contrary, not all the world's auctioneers are as honourable as they would like to have us believe. The large and important auction houses—the Sotheby's, Christie's, Phillips, and Bonhams of this world—may well be beyond all reproach but the same level of ethics does not necessarily apply across the board, and a degree of caution is essential. There are many small auction houses dotted across the land from whom you find you invariably buy at precisely the secret proxy bid you left by phone or fax. It may be argued (and it often is) that if you leave a bid of, say, $500 on a lot which the auctioneer has esti-

mated at $20-30, and you subsequently find that you have purchased the lot at (*mirabile dictu!*) exactly $500, then there is nothing really wrong since you declared that you were prepared to spend $500 in the first place.

Whether there is something wrong with that, or not, is for you to decide but it is as well to make a mental note that the outfit in question is not auctioning but is in effect horse trading, and this is not acceptable, in the author's view, since a great heap of consumer protection legislation which helps keep normal retailers honest is often negated by the fine print of disclaimers in the auctioneers' conditions of business. To give a specific example from a small auctioneer's catalogue which is typical of many, if not most, one of the clauses of their published Conditions of Sale states: 'All lots are sold as shown with all faults, imperfections and errors of descriptions and neither the auctioneers nor the vendors are responsible for errors of descriptions or for genuineness or authenticity of any Lot or for any fault or defect in same and no warranty is given or implied by the auctioneer or any vendor to any buyer in respect of any lot…etc. etc.'

It is as well to be aware, for instance, that an article is deemed by auctioneers to belong to the successful bidder upon the fall of the hammer, even though it may be hours or days (or even longer) before the successful bidder can effect payment and claim physical possession of the article. Good auctioneers take great care to try to prevent anything untoward happening. Lesser auctioneers may not be as fastidious and *caveat emptor* is not necessarily bad advice. The author is not the first man in history…nor will he be the last…who has paid for and cleared an auction lot of a machine which has all but halved in size since he last viewed it five minutes before the fall of the hammer, or which has had its lot number mysteriously switched for that of a comparable but lesser article in the same sale at some time between the fall of the hammer and clearing the lot. If the fields of botany and biology are replete with examples of predators and parasites, we ought hardly marvel at their prevalence in the human species. Of course, most people are too devoted and too busy to give all this much thought, which is a pity, really, because it benefits only those who give it all their thought. In the end, greatness sometimes rises to the surface, but in the collecting field as in physics, all too often it is merely the hot air which rises. And of course, it is as well to remember that fæces also floats.

Back to the point: as for determining values of old machines, there are some further comments to be made. What with modern communication technology, it might convincingly be argued that a machine ought to be worth the same anywhere in the world, but worth does not equate to monetary value, so that what one might reasonably expect to pay for a machine in the United States is by no means the same as one might have to pay for an identical machine in Great Britain or France. There is no logic in this, but it is a fact of life. This geographical factor may well be compounded by patriotism and national preferences: a rare German machine may well fetch immeasurably more in Germany than it does in the United States. On the other hand, the exact opposite may well apply to a common one.

To further complicate the issue, condition and completeness are absolutely vital factors affecting values. A perfect virtually unused machine, in its original carrying case with its original book of instructions and even at times its original receipt of purchase, will fetch a price which may well be out of all proportion to that of a lesser example. The same applies to machines which are pristine because they have recently been expertly overhauled—the labor costs of such a time-consuming task may well be considerable. Low serial numbers, rare mutants, uncommon variations of names, unusual specifications—all or any of these may throw the *cognoscenti* into a financial frenzy, resulting in prices which may bear little relationship to those of ordinary examples of the same make. The ubiquitous Mignon, for example, has been known to change hands at figures more readily associated with astro-physics, merely when it happens to have been manufactured in a color other than black.

To be of maximum value, a published price guide wherever possible ought to list a rough average spread, even though this often means disregarding the highest and lowest and quoting the spread of the rest. This is what I have more or less tried to do, with some exceptions, in quoting the values below.

At the end of the day, however, it must be borne in mind that the true value of a given machine is nothing more than what it is worth to an individual collector at a specific historical moment, and a market statistic for that make and model is established when the prospective buyer and vendor can convince each other to conclude a deal or when the auctioneer's hammer comes down on the highest bid. Collectors, therefore, are those who have the final word in determining values but understandably perhaps, this final word is often affluence and not passion. A purchase must first and foremost be a source of great satisfaction and boundless joy, and for this to apply suggests a certain degree of financial potential, because there is no joy to be had from a new acquisition if the kids won't eat next week as a result of the purchase. On the other hand, while a bulging pocket book may well guarantee its owner an instant collection, it will not necessarily guarantee him equivalent and ever-lasting satisfaction.

154. Elite (Courtesy of Bernard Williams)

ABEILLE

Little is known about this French typewriter introduced in 1904. It had visible typing, a keyboard of thirty keys, ribbon inking, and printed ninety characters by means of double shift. Despite the name, it is not to be confused with the totally different machine called Carissima which was marketed in England as the Bee. **(120)**

ACME

A conventional, four row front stroke machine designed by Zalmon G. Sholes. It was first promoted under the name Waterbury Standard Visible by the Zalmon G. Sholes Typewriter Company of Waterbury, Connecticut, in 1911 but its name was later changed to Acme.

Not a common machine, but not particularly interesting either (unless perhaps for residents of Waterbury) so if you come across one it should set you back very little indeed. A price tag in the high two or low three figures ought to be about right.

ADLER

This prominent German make dates back to 1896 when the bicycle manufacturer Heinrich Kleyer began building a thrust action machine under license. Patented by Wellington Parker Kidder in 1892, the design was produced by the Williams Manufacturing Co. in Canada under the name Empire **(157)** and in the United States under the name Wellington. For the first few years, the Adlerwerke continued to call their machine Empire, but dropped it in 1899 in favour of the name Adler, although the design and construction of the machine remained essentially unchanged. The first Adler series was given the designation Model 7, and in deference to its manufacturer's heritage, the machine incongruously included a bicycle wheel in its emblem, as well as the eponymous eagle. It was a highly successful and vigorously promoted machine and territorial battles between the original Empire and the later Adler interests inevitably ensued—battles which Adler eventually won.

The distinctive thrust action which was common to the Empire, Wellington, and Adler names (as well as to others such as the Noiseless and the Kanzler) was not actually Kidder's invention, having been patented by one Bernard Granville in 1889 and first used on an unsuccessful machine called Rapid. Nor was it Kidder's first typewriter design, which was the down stroke Franklin. However, the thrust action was an immensely successful concept which retained its popularity for decades after its introduction—the Adler continued to use it until the 1930s when it finally succumbed to the conventional type bar alternative.

The initial Adler licensing agreement bestowed upon the licensees unlimited rights to develop the design, which they did with a vengeance. The initial Model 7 design was subjected to minor sophistications such as tabulators, wide carriages, and a four row keyboard over the years, but the design principles remained essentially unchanged. A popular portable called Klein Adler, which was a Model 7 reduced in size and weight, was introduced in 1913. From the 1909 Model 11 onwards, still using the same mechanical principles, it was redesigned and presented a more angular and less aesthetically appealing profile.

Model changes did not automatically render predecessors obsolete and many models continued to be offered simultaneously. A 1926 catalog, for instance, lists the following (*inter alia*): Model 7, with various carriage lengths, tabulators, etc., Model 8 for interchangeable type which lift out as a unit, Model 11 with six characters on each type bar and double-double shift for a total of 180 characters, Model 16 with interchangeable type, Model 17 with four row keyboard offering 184 characters, Model 19 with four row mathematical and chemical keyboard of forty-six keys offering 138 characters, Model 25 as a large four row upright with billing and accounting attachments, tabulators etc.

The full-sized Adler was also marketed under the name Crown; the Klein-Adler portable under the names Adlerette, Adlerita, Adler Piccola, and Aigle.

Prices generally low, depicting lack of rarity. Depending on model and condition, some may even be worth up to $400 or so on a good day, but the rest should be much cheaper than that.

AEG

A conventional front stroke upright with four row keyboard introduced in 1921 by Allgemeinen Elektrizitäts-Gesellschaft of Berlin which was already marketing the very unconventional but enormously popular Mignon **(189, 190)**, which was also sold labelled AEG. Several similar models of the upright version were produced until the early 1930s when the name was dropped in favour of Olympia **(205)**. Value is low: you should get change out of $50 or so. **(110)**

AKTIV

Gustav Tietze AG introduced this primitive type wheel machine in Germany in 1913. The wheel was turned manually to select the letter, printing by pressing a lever; upper and lower case. Marketed in England under the name Active. **(111, 159 [top l])**

Expect to pay in the order of $500.

ALBUS

Austrian version of the Standard Folding, this lightweight aluminum folding three row portable machine was introduced in 1909 by Carl Engler of Vienna; manufactured until 1912, when the company was bought up by Clemens Müller of Dresden and the machine renamed Perkeo. Also marketed as Emka, Engler, Proteus, and Atlas.

Rarer than the Standard Folding **(249)** but worth roughly the same, except perhaps to an Austrian. $300-350 or so?

ALEXANDER

This conventional front stroke of small proportions was produced in limited numbers in 1914 by Alexander Typewriter Co. of New York. Invented by Jesse Alexander of Jackson County, Missouri.

ALLEN

One of the less successful of the many machines designed by the prolific Richard Uhlig, this three row double shift front stroke

155. Elliot Fisher (Courtesy of This Olde Office, Cathedral City, California)

156. Emerson (Courtesy of This Olde Office, Cathedral City, California)

machine was manufactured in 1918 by the Allen Typewriter Company of Allentown, Pennsylvania. Its eventual commercial failure was attributed to its three row keyboard; a sole example of the machine with a conventional four row keyboard survives but it was never marketed.

Three row version is worth some $200-300.

AMATA

An Austrian conventional upright with four row keyboard manufactured in 1923 by Maschinenfabrik J. von Petravic & Co. of Vienna. Approximately 1,000 were made.

Price in the low three figures may reflect rarity for an otherwise unremarkable product.

AMERICAN (1)

An 1893 patent granted to L. P. Valiquet of New York was assigned to the American Typewriter Co., which was already manufacturing a simple machine called American Visible **(114)**. On the American, rubber type was attached to a sector which swung on a vertical pivot beneath a corresponding letter index. A pointer selected the desired letter on the curved index; printing was by depression of a key on the left of the machine and inking was by roller. It was also marketed under the names Champignon, Globe **(165)**, Lion, and Sterling.

Poor examples are the more commonly encountered and are worth correspondingly less, but good ones can fetch $500-800. **(112)**

AMERICAN (2)

The same manufacturer introduced an up stroke machine with a carriage swinging up to reveal the typing. It was marketed as Armstrong in England, as Herald in France, and Europa in Germany; also labelled Eagle, Elgin, Favorit, Fleet, Mercantile, Pullman, and Surety. Production continued until 1915. $300-500 ought to be enough. **(113)**

AMERICAN FLYER

This toy typewriter using a sliding indicator geared to a type wheel was manufactured by American Flyer Mfg. Co. of Chicago, circa 1930.

$50-100, on a good day with a passing wind.

157. Empire

AMERICAN STANDARD

A blind, up stroke double keyboard machine manufactured in 1892 for a brief while before Remington successfully sued over the use of the word Standard, whereupon the name was changed to Jewett in honour of the president of the Company. According to information in the possession of the Dewitt Historical Society of Tompkins County, it was first produced in Parish, New York, by the Parish Manufacturing Co. which had been organized in 1886 by Lucien Crandall. By 1892, however, Crandall already had his hands full with his own problems and whether he retained an interest in the double keyboard machine is not known. The renamed American Standard, meanwhile, was manufactured by the Jewett Typewriter Co. of Des Moines, Iowa, and sold under the name Jewett.

$1,000-1,500 ought to be more than enough.

AMERICAN VISIBLE

A primitive linear index machine manufactured by the American Typewriter Company from 1893. Type was on a rubber strip with a marker attached and the desired letter on it was located by placing the marker over the corresponding letter on the front of the machine which was then pressed down for printing to be performed, inking by pad.

A second model introduced later presented a modified marker and a rigid index, printing now performed by depressing separate levers on the side of the machine.

Reckon on spending anywhere from $1,000-1,500 on, depending on condition. **(114)**

ANDERSON

G. K. Anderson of Tennessee patented a shorthand machine from 1889 onwards. Fourteen keys were used to print phonetically on paper tape by means of the simultaneous depression of one or more keys. It was manufactured in some numbers as the Dictatype, by the Dictatype Shorthand Machine Co. of Philadelphia.

$50 or so.

ANNELL

Introduced on a mail order basis in 1922 by Annell Typewriter Co. of Chicago, this conventional front stroke with four row

keyboard was the same as the 1914 Woodstock Model 4. $100 ought to leave you with change in your pocket.

ARCHO

A German machine with a linear thrust action of Empire-Wellington-Adler inspiration, offered as a three row model from 1920 with a four row version from 1935. Manufactured first by Winterling & Pfahl, later by Inh. Carl Winterling, numerous models were made, including one which typed letters of the alphabet as well as musical notes, one which typed only music, and a portable slim enough to be carried in a briefcase yet sturdy and stable, the introduction of which was interrupted by the Second World War.

Price around $200-300.

ARLINGTON

One of the machines invented by the prolific Richard Uhlig, this visible three row keyboard type bar machine with double shift was manufactured in 1914 and named after the town in which the inventor was living at the time.

Not often encountered, but not particularly desirable unless you come from the town. $100, perhaps?

ATLAS

The Atlas Typewriter Company of New York manufactured a three row keyboard machine with double shift and an oblique front stroke action in 1915. Invented by Richard Uhlig, who was responsible for some fifty different typewriter designs.

Relatively rare, but not expensive for all that. Three figures ought to do it.

AUTOMATIC

Despite having nothing whatsoever to do with automation in any form, this misnamed machine sported a three row key-

158. Enigma (Courtesy of Sotheby's, London)

159. Eureka **(top c)**, Active **(top l)**, Babycyl **(top r)**, Simplex **(centre l & lower r)**, Dollar **(lower l)** (Courtesy of Christie's South Kensington, London)

board, type bars 1.5 inches long and differential spacing. It was a compact up stroke instrument, printing upper case only, with inking by pads. It was patented by E. M. Hamilton and assigned to the Hamilton Type-Writer Co., the patent application being filed in 1884 and granted three years later. A small number was manufactured but few have survived. Also known under the name Hamilton.

Very, very rare, hence a price tag in the $10,000-15,000 range.

AVISO

A front stroke machine with three row keyboard briefly manufactured in 1923 by Otto Schefter of Berlin.

Cheapie.

BABYCYL

One of the better tin plate toys of small dimensions, typing by means of a proportionately large type wheel prominently positioned on top of the machine.

Three hundred dollars or so, perhaps a little more. **(115, 159 [top r])**

BAR-LET

The Barlock Company (see below) produced this line of small portables in England offering three and four row keyboard models from the 1930s onwards. These were the same machines as those previously introduced in Germany under the name Mitex.

Bar-Lets may fetch three figures, if condition is good. **(116 [l])**

BAR-LOCK

An inventive watchmaker by the name of Charles Spiro of New York was responsible for this fine machine which he patented in 1889 and launched on the market the same year. Made by Columbia Typewriter Manufacturing Co. which had already produced a small type wheel device so beloved by collectors today (called, appropriately, Columbia), the Bar-Lock nevertheless was the machine which made Spiro his fortune from the hundreds of thousands sold around the world over the course of several decades following its introduction.

The Bar-Lock was a down stroke design with a full keyboard made originally with a distinctive front cover around the arc of vertical type bars; this was soon replaced by a plainer one in pressed metal painted black which was retained until the Model Fifteen, introduced in 1911 as an oblique front stroke, making typing more easily visible to the operator. The name of the machines was derived from the small arc of pins which served as guides to lock the type bars into position for correct alignment and to prevent jamming.

Model 12 in 1907 was the first Bar-Lock to offer a standard four row keyboard and shift, and the name was simplified to Columbia.

Meanwhile, machines were being marketed under the name Columbia Bar-Lock. Those built in England had been called Royal Bar-Lock and this continued until Model 15 whereupon, to complete the confusion, the manufacturers reverted simply to the original name Bar-Lock.

The US manufacturing company sold out completely to the Barlock Typewriter Company Ltd of London in 1914 and production of the machine—not merely assembly of the US parts—was continued on the other side of the Atlantic after the War. The first product was a four row front stroke machine of ungainly profile marketed as the Model 16 in 1921. Minor model changes followed. Model 17 was manufactured from 1925 to 1929; ownership of the company meanwhile passed to the manufacturing firm of John Jardine Ltd and the now All British Bar-Lock was manufactured in Nottingham by Barlock (1925) Ltd. Further model and name changes followed until the 1950s when the machine was finally withdrawn.

A line of smaller machines sold under the name Bar-Let included models with three and four row keyboards and was produced by the company from the 1930s onwards. These were the same as machines previously made in Germany and sold under the name Mitex.

The first model is likely to be up in the $4,000-5,000 bracket, and worth it; subsequent models become progressively and

160. Fitch, in original cabinet (Courtesy of Bernard Williams)

considerably cheaper right down to the conventional machines, which are worth very little. **(90, 116 [l], 117 [r], 118, 119)**

BARR

Conventional oblique front stroke machine invented by John Barr in 1926 and produced by Barr-Morse Corporation of Ithaca, New York, until 1934 and by Barr-Morse Co. Ltd of Montreal, Canada, after that.

$100-150 ought to do it.

BAVARIA

A small front stroke portable with three row keyboard and double shift was introduced in 1921 and manufactured for a short time, first by Bavaria-Schreibmaschinenfabrik Gebr. Siegel, later called Ria Büromaschinenfabrik Gebr. Siegel & Co. GmbH.

Ho-hum! Bavarians might pay…what?…$50 bucks for it? No one else would.

BENNETT

This younger brother of the miniature Junior appeared in 1910, made in the Elliot-Fisher factory and distributed by the Bennett Typewriter Co. of Harrisburg, Pennsylvania, although a New York address appears on some machines. It was virtually identical to the Junior, the principle technical difference being the substitution of a ribbon for the earlier model's roller inking. See entry under Junior. **(1)**

Cute and good value around the $400-500 mark. **(121)**

BERWIN SUPERIOR

Manufactured in the 1940s, this toy type wheel machine had a painted keyboard. $30…$40?

BING

A lightweight portable of tin-plate construction was manufactured in Germany by Bingwerke AG in the 1920s with a four row keyboard and oblique front stroke type bar movement. It had roller inking in the first model, ribbon in the second. The company also manufactured tin-plate phonographs, toys, etc. $200-300 ought to be enough. **(122)**

BLICKENSDERFER

George C. Blickensderfer patented his first type wheel three row keyboard machine in 1889 and it was produced by the Blickensderfer Manufacturing Company of Stamford, Connecticut and marketed as the Model 5 from 1893 onwards. It was an instant success and sold in vast numbers through successive models. A Model 3 is known to have existed but it and other early versions were largely developmental.

The principal features of the Model 5 and its later refinements were the use of interchangeable type wheels, roller inking, and Ideal three row keyboard. It was a lightweight machine and small enough to be genuinely portable, yet an aluminum version of the Model 5, even lighter in weight, was offered as a Model 6 and a range of Featherweight Blicks was subsequently marketed in large numbers. Model no. 7 featured a space bar which extended around the sides of the keyboard and a scale above the platen, Model 8 introduced in 1908 added a tabulator, and Model 9 sported a folding inking roller support.

The versatility of the design lent itself to a multitude of specialized uses: Oriental models featured carriages which could travel either left-to-right or right-to-left at the will of the typist, tabulators and long platens were offered and the machine was even marketed as an electric in 1902. This was a revolutionary concept at the time, and inevitably it failed as a commercial venture, although some 150 electrics are believed to have been manufactured. A second attempt at launching the electric model just prior to Blickensderfer's death met a similarly dismal fate.

1902 also saw the introduction of an index machine called Niagara using Blickensderfer components but substituting a circular index for the keyboard, and a musical typewriter based on the Blickensderfer and called Noco-Blick was made in Germany in 1910. Universal keyboards became optional in the same year.

A type bar model of front stroke design with a four row keyboard called Blick Bar had only just begun to establish itself in

161. Fox (Courtesy of This Olde Office, Cathedral City, California)

162. Fox 23 (Courtesy of This Olde Office, Cathedral City, California)

1916 when Blickensderfer's death the following year threw the company's affairs into confusion. The famous type wheel models were discontinued, the three row front stroke portable Blick Ninety **(10 [r])** was unsuccessfully launched in 1919, becoming the Roberts Ninety in 1922, and the Blick Bar was briefly sold as the Harry A. Smith, named after the entrepreneur who bought it up following Blickensderfer's death.

A son of the former Blickensderfer sales manager recalled hearing his father reminiscing about a voice command typewriter which was allegedly in advanced stages of development in the Blickensderfer factory between 1910 or 1911, and 1917. The operator could dictate directly into the machine which typed the sounds phonetically. Blickensderfer is alleged to have toyed with marketing the invention but he died before the scheme was realised.

In France it was marketed under the name Dactyle **(146)**.

In Britain, the Blick Typewriter Co. of Cheapside, London, launched a conventional front stroke upright with four row keyboard as late as 1924 under the names British and British Blick, and in the United States, a final attempt at giving the type wheel design the kiss of life was made by Remington in 1928 with the introduction of the machine under the names Rem-Blick and Baby Rem. By this time, however, *rigor mortis* had set in and the corpse proved beyond resurrection.

Wonderful machines, but of most models too many have survived. Condition is critical: very good examples may even fetch up to $500 or $600. Oriental model is relatively rare and consequently may fetch more. The same for Niagara, and if an Electric comes up, think five figures and then work out how high! **(10 [r], 49, 63, 123, 124, 125, 126, 146)**

BOSTON

D. E. Kempster of Boston patented a type wheel machine in 1886 with improvements filed the following year and the Boston Typewriter Co. began manufacture in 1888. The design had a curved letter index from which the desired character was selected by means of a lever which turned a quite substantial type wheel mounted vertically above the paper. A rare machine of which few have survived, it is often confused with a different instrument altogether, more commonly called World, which was also marketed on a small scale under the name Boston. The inventor had previously been involved with W. H. Gilman in patenting a cord-operated type wheel design.

If you get it for less than $10,000 or so, you risk being charged with theft! **(127)**

BRETT

Several printing telegraphs were designed by Jacob Brett in England and Royal E. House in the United States from 1845 onwards and although they appear to have made some progress with their designs, these were eventually abandoned. The exact relationship and the circumstances of the co-operation between these two men has never been fully understood. Brett's British patents pre-date those of House but neither man's patents state them to be 'on communication from abroad' as was common practice when patents were applied for on behalf of overseas inventors.

The 1845 patent for a printing telegraph used a rotating cylinder on the transmitter synchronised with a similarly rotating type wheel on the receiver. Depression of a key on a piano keyboard running the length of the transmitting cylinder caused contact to be made with its corresponding pin, thereby stopping the cylinder at one end of the circuit, and the type wheel at the other end, at the same letter. A further patent was granted in 1848, and Brett's instrument was used for a time both in England and in Europe but was eventually dropped. Royal E. House, on the other hand, continued promoting and improving the design in the States and eventually enjoyed considerable success.

Too rare to quote a value, but that does not necessarily mean it is priceless.

BROOKS

A down stroke machine with three row keyboard and double shift, with the type bars positioned vertically behind the platen, was patented in 1885 by Byron Brooks, a versatile inventor with considerable experience in the field, having worked for a time on the development of the Sholes and Glidden. Among the innovative features of the Brooks was the use of a device which automatically lifted the ribbon to make the typing visible, and a vibrator which moved the ribbon laterally to use its full width. It was manufactured by the Union Writing Machine Co. of New York; W. P. Hatch of book typewriter fame was for a time in charge of sales.

Five figures.

BUCKNER LINO-TYPEWRITER

A Smith Premier No 1, modified for linotyping, was marketed under this name. The modification entailed removal of the lower row of keys and space bar and replacement by a blanking plate bearing the legend Buckner Lino-Typewriter Co.,

163. Geniatus (Courtesy of This Olde Office, Cathedral City, California)

164. Gladstone (Courtesy of Musée National des Techniques C.N.A.M., Paris)

Berkeley, California. The space bar was replaced by one placed and shaped like the space key on a linotype machine.

Rare, but $1,000-2,000 should be quite enough.

BURNETT

This oblique front stroke machine with four row keyboard first appeared in 1905, and in slightly different dress as the Triumph Visible in 1907. With its type bars enclosed within an unsightly cowl, it was marketed briefly by the Burnett Typewriter Co. of Chicago and retailed by Sears Roebuck. Harry A. Smith, so graphically described as 'the trade's paramount decalcomaniac' by the late Dan Post, appears to have had a hand in it, no doubt in the disposal of unsold stock.[87]

$5,000-10,000.

BURNS

An up stroke machine with a double keyboard and a more than casual resemblance to the early Smith Premiers (amongst others) was patented in the United States by Frank Burns of Westfield, New York, in 1889. The Burns Typewriter Co. was formed the following year to market the instrument. Its features included such innovations as variable line spacing, back spacing, and interchangeable platens but few machines were made and the company turned its efforts more profitably to making steel type for its more successful competitors.

The inventor had previously begun experimenting with a linear index machine with rubber type in 1888 but had dropped the project in favour of his up stroke design.

Rare, hence up to ten grand. **(128)**

BURROUGHS

This major manufacturer of accounting machines briefly entered the typewriter market in the early 1930s with a conventional upright which they later fitted with electric carriage advance and return. The venture was short-lived. **(129, 130)**

$200 to $300, or so?

CAHILL

An early attempt at an electric typewriter was patented in 1893 *et seq* by Dr. Thaddeus Cahill of Washington, D. C., and some forty units are reported to have been manufactured. The machine, which bore a strong resemblance to a Remington No 2, featured electrical operation of all moving parts including carriage and type bars.

Too rare to quote.

CALIGRAPH

A fine machine, and one of the big names in the early years of the typewriter's introduction, the Caligraph was invented by George Washington Newton Yost and Franz Wagner, the design covered by a patent filed by the former in 1880 and granted in 1883. Yost's life-long love affair with the typewriter began as far back as the early 1870s when he first associated himself with Sholes, Glidden, and Densmore in the promotion of the infant Type Writer. He was later to break away after Remington assumed full control. Production of his Caligraph by the American Writing Machine Co. began in 1881 with a model which typed only capital letters, to be followed a year later by the No 2 which printed both upper and lower case. It was reported that 13,000 Caligraphs had been sold up to the end of 1886.[46]

The Caligraph was an up stroke machine, with the type bars in the conventional circular basket under the platen. Its uniquely elongated profile, however, was the result of the type levers being pivoted at the near rather than at the far end of the machine, the ends of the levers concealed beneath a flat metal plate which was subsequently claimed to have been purposely designed to permit the typist to rest his or her hands on the machine during typing. From upper and lower case Model 2 onwards, the machine sported at first 'full' keyboards with white lower case keys in the centre surrounded by black upper case keys, and then 'double' keyboards from Model 5. It became the 'New Century Caligraph' from 1900 on, later simply 'New Century,' the elongated profile now abandoned but the keyboard retained. Production ceased in 1906 by which time it was controlled by the Union Typewriter Co.

Difficult to call, because there are so very many models, but good early machines may make up to $1,500 or so, later models very much less. **(65, 131, 199)**

CARDINAL

A conventional upright which was to have been manufactured in 1914 but which appeared only after the First World War. The watchmaking factory D. Furtwängler Söhne AG of Furtwangen, Baden, produced the machine until 1926.

You should get change from $200-300.

CARISSIMA

A compact type wheel machine of late manufacture, this bakelite anachronism was made in 1934 by Th. Knaur-Hübel & Denk of Leipzig, later Kurt Strangfeld of Berlin. An indicator on the front of the machine selected the desired letter on a linear index and type wheel, printing by ribbon. Marketed in England as 'The Bee.'

$500-750 should be plenty.

CARMEN

Conventional three row front stroke machine introduced in Germany in 1920 by Carmenwerk AG of Stuttgart.

$200 or so.

CASH

Arthur Wise Cash of Hartford, Connecticut, patented a down stroke machine in 1887 with a four row keyboard and ribbon inking which typed on paper held in a flat bed. It was manufactured first by him, and then from 1896 by the Typograph Co. when the name of the machine was changed to Typograph. Very, very rare machine, this one, so if you can, write a blank cheque. If not, think five figures. **(133)**

CELTIC

The most interesting product of this French manufacturer was a three row portable with a down stroke action, the type bars folding back when not in use to give the machine a lower profile. This model was first introduced in 1923, but the company had already been in the market for a decade (with interruptions) with a conventional front stroke four row upright introduced just before the War and then produced from 1919 by Société des Moteurs Salmson, later by S. A. Celtic.

Cute and relatively unusual but not particularly costly, except perhaps in France. Nevertheless, budget for a three figure price tag.

165. Globe (Courtesy of Bernard Williams)

166. Graphic (Courtesy of Bernard Williams)

CENTURY (1)

Thomas Hall, inventor of the popular index machine which bears his name and which he patented in 1881, was responsible for two other designs: an earlier type bar machine, and the later Century which was clearly inspired by the success of his index instrument. In fact, the Century's index and indicator are quite unashamedly from the Hall but with the indicator geared to a type sleeve in a stationary sub-frame with the carriage moving as on more conventional designs. Inking by roller. Rare and desirable, hence four figures.

CENTURY (2)

The American Writing Machine Co., one of the members of the Union Typewriter Co., produced a front stroke upright with three row keyboard and double shift by this name. Designed in 1914 by Frank Sholes, one of the sons of Christopher Latham. A similar machine was marketed as the Remington Junior.

$300-400 or so should do it.

CHAMPION

Last of a trio of comparable index machines (the other two are Peoples and Pearl) invented by Carl Sjoberg of Brooklyn, New York, the precise chronology of which has presented typewriter historians with some interesting posers. The Champion is of the type wheel class, with shift keys for upper and lower case, the type wheel geared to a selector over a curved letter index. The inking system was by means of a wide spool-to-spool ribbon, which represents one of the principal differences between this machine and its two above-mentioned siblings. It was protected by a patent filed 2/9/1892 and granted 7/3/1893 and was manufactured by Garvin Machine Co. of New York until the factory was destroyed by fire in 1898.

Expect prices around $1,500. **(134 [I])**

CHICAGO

Horizontal type sleeve machine with three row keyboard previously sold as the Munson, the name being changed in 1898 when the manufacturer became the Chicago Writing Machine Co. A slightly modified Model 3 sporting minor changes to ribbon width and location, carriage, etc. was introduced in 1903 and production continued until 1912 when the manufacturer became the Galesburg Writing Machine Co. and the name of the machine changed, predictably, to Galesburg.

Also marketed under the names Baltimore, Competitor, Conover, Draper, Ohio, Standard, and Yale.

Condition will determine final price in the general region of $1,000+ for the Chicago. *Cognoscenti* may feel tempted to pay a little more for some of the rarer titles. **(135, 140)**

CHINESE

Commercial Press Works of China produced a typewriter invented by Shu Chen Tung, one of their engineers, which overcame the complexities of typing 2,000 or so characters in the Chinese language by using a mobile platen which could be

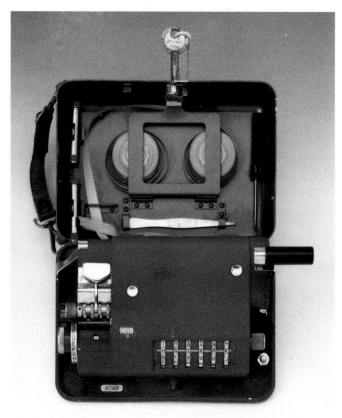

moved in all directions over the surface of square trays containing the type. With ribbon inking, it dates from the early 20th century.

Think of a price, and then haggle.

CIPHER

Frederick Sedgwick invented this machine during the First World War and it was manufactured by the International Cipherwriting Co. of Chicago in 1919. It was based upon a Hammond typewriter, scrambling the relationship between keyboard and type shuttle.

Rare but not necessarily valuable as a result. Base yourself on the comparable Hammond price and add a little for the novelty.

CLAVIER IMPRIMEUR

Final design by French typewriter pioneer Pierre Foucauld and arguably the first machine ever to be manufactured, beating the Typograph for this honour (if indeed it did) by a hair's breadth. The reason for the doubt is that there are no definitive records for the actual dates of production, but on the strength of publication of specifications the Clavier Imprimeur wins the title by a year.

Foucauld's involvement in the development of the writing machine dates back to a fascinating 1839 invention which he called Raphigraphe. It was not successful, nor was a modified version completed some years later. The experience gained was put to good use in the 1849 Clavier Imprimeur which was of radial plunger design consisting of two parallel rows of thirty plungers each, radiating outwards from the common printing point on a flat paper table beneath. The row of keys furthest from the operator was vertical while the nearer one was inclined inwards. It used carbon paper, had a bell to indicate end of line, and a space bar between the two rows of keys. It was optionally offered for perforating or embossing for the use of the blind.

The machine was completed in 1849 and was already being acclaimed the following year. In 1851 it was awarded a gold medal at the London Exhibition by which time manufacture was apparently underway. It was still being sold and used at the time of the following London Exhibition in 1862.

If you find one, offer your Cadillac, and if that won't do it, try the house…not instead of the Cadillac, but as well as! **(16)**

COFFMAN

A miniature linear index machine with type on an endless rubber belt was patented by G. W. Coffman of Garden City Kansas in 1902. It was manufactured in small numbers, retailing for $10.

Not ten dollars any more. Very rare—if you buy one for less than $5,000-10,000 or so you have done well.

COLUMBIA

A prolific New York watchmaker called Charles Spiro was the inventor of this fine little machine so beloved by collectors. Patented in 1885 and marketed by the Columbia Type Writer Co. of New York, it printed by means of a vertical type wheel turned by a substantial knob and geared to an indicator which selected letters on a circular index. Printing was by means of pressing down on the knob, a motion which simultaneously moved the platen to accept the fresh impression, and by an ingenious linkage behind the type wheel, actually moved it the correct amount for the width of the letter being typed, thereby providing one of the earliest manufactured examples of differential spacing.

Inking was by means of a small circular pad beneath the type wheel which was displaced laterally as the type wheel was lowered. Regrettably this flimsy device was so successfully designed for easy removal from the machine that it is now often missing on surviving examples.

Three models were manufactured. The first offered only upper case; this was replaced by a second model in which upper and lower case were on two separate type wheels mounted parallel to each other, which itself was almost immediately superseded by the machine in its final form, with a single type wheel printing both upper and lower cases. This model was

169. Hammond (Courtesy of Christie's South
Kensington, London)

labelled 'Improved No 2', and in contrast to its two predecessors, enjoyed considerable commercial success. Serials numbers of upper case only, and double wheel, models are low, indicating only modest production: a double wheel example previously in the author's collection bore serial number 314. Improved No 2s run into four figures.

'It is to the Pen, what the Sewing Machine is to the Needle,' claimed a contemporary leaflet issued by the London agents. 'Highest Awards at the 1885 Exhibitions at London, New Orleans, and New York…The speed attainable is 40 words per minute, more than any other portable machine, thus equalling the best key machines…the best and cheapest upper and lower case Type Writer in the market.' And at £6. 6. 0 (including 'handsome polished wood box, with lock and key, and flask of ink') who would argue with that?

£6. 6. 0 was six guineas, or six pounds six shillings in old money, or £6.30 in modern currency. If you find one at that price, ask not to be woken. More likely that not, however, you will need $4,000-6,000 these days. The upper and lower case model 2 is the most common, the double wheel model is the rarest. **(9, 21, 48, 73, 137)**

COMMERCIAL

Conventional upright manufactured from 1914 by Commercial Schreibmaschinenfabrik K. Fr. Kührt of Nürnberg. Numerous models. Also marketed under a multitude of names including Arpha, Atlantia, Berolina, Condor, Constanta, Esko, Hansa, Heros, Industrie, Mafra, Media, Mepas, Mercedesia-Record, Orbis, Orplid, Progress, Resko, Thüringen, Universal, etc. The design featured easily interchangeable type baskets for different languages and uses.

Cheap. $50, perhaps? Not to be confused with the following machine.

COMMERCIAL VISIBLE

One of the prolific Richard Uhlig's fifty-or-so designs, this machine enjoyed only limited success. It used a type wheel with three row keyboard and double shift and was protected by a patent filed in 1897, with the first machines made the following year. Also marketed by a New York department store under the name Fountain. The manufacturing company was originally Visible Typewriter Co., later Commercial Visible Typewriter Co., both of New York.

$2,000-2,500 should be plenty. **(117 [c], 138, 139)**

CONDÉ.

An oblique front stroke machine with a difference, this four row keyboard design by Samuel Condé of Illinois patented in 1894 featured a platen raised high enough at the rear of the machine to make typing visible to the operator. Very few were manufactured and fewer still, if any, have survived.

CONSTANÇON

A Swiss machine invented in 1910 by Maurice Constançon for the use of the blind, this invention allowed the operator to emboss braille cells on both sides of a page using the spaces between the lines. Two models were made.

A few hundred dollars should secure it.

CONTIN

A French conventional upright first introduced in 1922 by Établissements Continsouza of Paris. Models A to E appeared between the Wars. It was sold extensively in Poland, under the name F K.

Cheap. A good one ought to be in the high two or low three figures.

CONTINENTAL

A conventional German upright produced in large numbers and models for several decades. First introduced in 1904 by Wanderer-Werke vorm Winklhofer & Jaenicke A. G., Siegmar-Schönau.

Also cheap. Very good ones have been known to make into the low three figures—just! **(100)**

CORONA

A small lightweight portable featuring a carriage which pivoted forward and over the keyboard when not in use, making it compact and easy to carry. The design was the work of Frank

170. Hammond Multiplex (Courtesy of This Olde Office, Cathedral City, California)

S. Rose of New York and first appeared as the Standard Folding **(249)** until 1912 when the name was changed to Corona, and two years later the manufacturing company also changed its name to Corona Typewriter Co. During the course of the change-over, machines were made with both names appearing: 'Corona (Standard Folding Typewriter) Manufactured by Standard Folding Typewriter Co.' A conventional non-folding model was introduced in 1923 but the folding model continued to be manufactured until World War II and it remains to this day one of the most ubiquitous of all old typewriters. Production figures extend to over 700,000 by 1941.

Manufactured in Germany under the name Erika by Seidel & Naumann A. G. of Dresden. Atlas, Corona Piccola, Franconia, Improved Corona, and Piccola were all names under which it was sold in different parts of the world.

In 1926 the Corona Typewriter Co. merged with L. C. Smith Typewriter Co. of Syracuse and the well known Smith-Corona products were the result. A final merger with Marchant Calculators in 1958 opened up the data processing market.

$200 or so, but only for very good and complete examples. **(141, 142)**

CRANDALL

The fine type sleeve machine usually associated with this name is only one of the many contributions made by typewriter pioneer Lucien Stephen Crandall. His inventive career began with an 1875 patent for an unusual up stroke design with a keyboard consisting of a curved row of eight keys, with a similar number of type bars each containing six characters. It has been suggested that the machine was designed for the use of the blind and this would appear to make good sense, but the project never left the drawing board.[38]

Shortly afterwards, Crandall and James Bartlett Hammond, who was soon to lend his name to one of the greatest makes of all time, were reported to be involved in a conflict over control of Pratt's patents. Hammond emerged the winner and Crandall set out to produce an original design which he could patent himself.

The result was a machine with a straight three row keyboard controlling a long type sleeve of small diameter, with characters in six rows, which was located obliquely above the platen. It was ribbon inked. The design was granted a British patent in 1879 and a US patent two years later. Manufacture was undertaken in 1881 first in Syracuse and later, from 1887, in Groton New York by the Crandall Machine Co. Several thousand units are claimed to have been made, but few have survived. These were still early days in typewriter manufacture; however, according to a contemporary source, success soon followed and brought 'unprecedented prosperity to Groton.'[38] Meanwhile, a redesigned version called 'New Model' had appeared in 1886. This instrument retained the type sleeve principle but the keyboard was modified to a curved two row design with a vertical rather than horizontal profile. This model was made in considerably larger numbers—an example previously in the author's collection had a serial number over 17,000—but not many appear to have survived.

A third change resulted in the Universal Crandall of 1893 with a somewhat modified appearance and straight three row keyboard, while a fourth model put in a brief appearance some years later.

A totally different design with elements of Crandall's 1879 creation was patented in 1895 and labelled Improved Crandall. It had a straight three row keyboard controlling twelve horizontal type bars, each with six characters on the undersides. Upon depression of a key the type bars moved horizontally forward to the selected character with a hammer striking from above producing the impression. Having used ribbons on previous machines Crandall now tried an inking roller but reverted to ribbon on a later modification. It was also marketed in England as the Cosmopolitan.

Despite reports of the continued success of the Crandall enterprise the company gradually declined and even closed for a time according to contemporary accounts (*49*). It also tried to save itself by diversifying into such activities as special tool production and even, following the German example, into bicycle manufacture with its 'Lehigh' brand of 1896. Both the Universal Crandall Typewriter No 3 and the Lehigh bicycle

171. Hammond VariTyper

172. Hammonia (Courtesy of Christie's South Kensington, London)

173. Malling Hansen's Skrivekugle (Courtesy of Christie's South Kensington, London)

174 (a & b). Malling Hansen's Skrivekugle—details
(Courtesy of Christie's South Kensington,
London)

began to appear together in the same advertisements and from these and other contemporary sources it would now appear that for some time both the type sleeve and the type bar machines were manufactured simultaneously and the advent of the latter did not immediately and automatically herald the demise of the former. Typewriter production may also have continued for longer than once believed, and 1899 sources[38] were still speaking of Crandall manufacture in the present tense. A Crandall machine bearing the name Fennia is known.

Expect to pay up to $8,000-10,000. (69, 143, 261)

CRARY

An interesting design for a book typewriter was patented by J. M. Crary of New Jersey in 1892 and produced in small numbers. The down stroke type bar machine had a keyboard of three concentric circles above a type basket which travelled along rails front and rear, the type levers striking downwards onto the flat pages.

Too rare to quote. Just pay the asking price and run.

CROWN (1)

An ingenious if complex solution to providing upper and lower case on type bars in a circular formation was patented in 1887 by A. G. Donnelly of New York and promoted by the Crown Typewriter Manufacturing Co. In this design the lower ends of the type bars were geared to a ring so that a slight twist one way or the other selected the desired case on the entire circle,

which was mounted obliquely above the platen. The whole thing resembled a crown from which the machine derived its name. It apparently retailed for $20 although none appears to have survived. Also reportedly marketed under the name Donnelly.

Price? Same as Crary, above.

CROWN (2)

One of the pioneers of typewriter history with credentials dating back to the earliest days of Remington's developmental work on Sholes and Gliddens, Byron A. Brooks was responsible for numerous patents from 1883 onwards, including several using type wheels, although the machine known as the Crown was not manufactured until the 1890s. An indicator geared to the type wheel selected letters from a straight index. Roller inking.

$10,000-15,000, without a whimper. (144)

CULEMA

A front stroke three row keyboard machine with double shift manufactured in Germany by Gebr Lehmann from 1919, with different manufacturers and slight design changes in succeeding years.

$100-150 or so should seal it.

DAISY POINT WRITER

An American called Joel Smith invented a machine for the blind as far back as 1865, with periodic improvements over the following fifteen years. Six keys positioned like the petals of a flower gave the design its name.

$300-500 or so.

DARLING

One of several primitive machines manufactured by Robert Ingersoll and Bros of New York, better known for their range of cheap watches, toys, etc. The machines ranged from a hand held type wheel model to ones which could be clamped to table tops.

Don't be shocked by sophisticated prices, all the way up to $4,000 or so. (147)

DART

An 1890 patent was granted to L. Dart and assigned to the Type-Writing Machine Co. of Connecticut for a sign printing machine on wheels with a type wheel 18 inches in diameter which moved a space to the right after each impression. A

175. Horton (Courtesy of Tom Fitzgerald)

knob on top of the instrument geared to the type wheel selected the desired letter. Pressing it down printed the letter and moved the device to the next space. It was roller inked. Rare but not necessarily priceless. May reach the high three figures.

DATTILO-MUSICOGRAFO

An Italian, Andrea Ferretto, was the inventor of a musical typewriter in 1921 which was manufactured for several years.
The same as Dart, above.

DAUGHERTY

The first full front stroke machine to be manufactured offering truly visible typing (the Horton was oblique) was invented by James Daugherty and granted a US patent in 1891. It had a four row keyboard with the platen raised vertically above the level of the horizontal type bars. Ribbon inking. It was designed so that the keyboard and the type bars could be removed as a single component. First manufactured in the Crandall factory in New York and from 1890 in Pennsylvania by the Daugherty Typewriter Company. It was renamed Pittsburg in 1898.
$2,000 or so ought to seal it. **(148)**

DAW & TAIT

A rare and desirable machine with a circular two row keyboard above a corresponding circle of type bars which printed by down stroke on paper on a cylindrical platen. Thomas George Daw and Hilder Daw of Sevenoaks in Kent were printers by trade and were granted a number of British patents in 1884 and 1885 protecting their design which was sophisticated enough to boast differential spacing but used carbon paper or ribbon positioned between platen and page to produce the impression on the underside of the paper. Several alternative arrangements of platens were protected in the patents, including a flat paper table which advanced along a spirally cut Archimedian screw, and later, a cylindrical platen of relatively large diameter with the typing around rather than along the cylinder.

Exceedingly rare. Comfortably into five figures. Anything less is classifiable as larceny **(149 a & b)**.

DAYTON

A front stroke four row portable manufactured briefly in 1924 by Dayton Portable Typewriter Co. of Dayton, Ohio, which was announced with considerable fanfare the previous year: 'We use the same methods in production and sale of the Dayton as Henry Ford does in making his cars.' Maybe, but if the latter had enjoyed the same measure of success as this typewriter, we would still be riding horses.
Not an exciting machine. Low three figures should be plenty.

DEA

A conventional upright originally called Union but the name was already protected by Union Typewriter Co. It was manufactured in Germany from 1908, first by A. G. vorm Gustav Krebs, later from 1917 by Concordia Maschinenbau A. G. Also marketed under the names Radio and Concordia. Production continued until 1932.
$50 ought to buy the best there is in the world.

DEMOUNTABLE

An American machine designed for easy disassembly into its basic components was manufactured in 1921 by the Rex Typewriter Co. which had been making various typewriters since 1911 under the name Harris Typewriter Co. before changing to Rex in 1914. The company changed its name yet again to Demountable Typewriter Co. in 1923, continuing to trade until 1936. The Demountable was a conventional upright with four row keyboard.
Condition might raise the price to $200-300. **(101)**

DENNIS DUPLEX

An interesting if misconceived attempt at increasing typing speed by printing bigraphs through simultaneous use of two keys, this up stroke type bar machine manufactured by the Duplex Typewriter Co. of Des Moines was protected by a patent granted to A. S. Dennis in 1895. This was itself for improvements to one granted in 1884 in the name of Henry Orpen of Missouri protecting modifications to Remingtons and Caligraphs to permit two keys to be used simultaneously.
On Dennis' invention, each hand operated its own half of the keyboard of some 100 keys with the type in two semi-circular baskets separated by a one space gap between their printing points. Inking was first by pad and later, when the machine was simply called Duplex, by ribbon.

176. Hughes Typograph (Courtesy of Dan Post Archives)

177. Ideal (Courtesy of Bernard Williams)

It was also labelled Germania-Duplex. The Jewett was a later development of the same design.

$2,000+, or so, ought to be sufficient.

DENSMORE

Type Writer pioneers Amos and Emmett Densmore, together with step son Walter Barron and the talented Franz X. Wagner (who was later to invent the great Underwood) teamed up to design this up stroke machine with a four row keyboard and such novel features as the use of bearings on type bar pivots and carriage hardware. Their company, appropriately named Densmore & Densmore, launched the machine in 1891, with successive model changes until 1907 when a Model Five was introduced. Shortly thereafter, the company was absorbed into the typewriter trust called Union Typewriter Company and phased out.

Not rare, so condition is likely to be important. $500-800 should do it. **(95)**

DIAL

A tinplate toy typewriter produced by Louis Marx in the 1930s with type wheel and roller inking. There were several models, including one labelled De-Luxe which had a mock keyboard painted on the front.

$50? $60?

DIAMANT

A front stroke machine with three row keyboard and double shift produced by Diamant Schreibmaschinen GmbH from 1922 to 1926. It was retailed in England as the Diamond **(150 [1])**.[1]

$100-150.

DIAMOND (2)

A conventional front stroke with four row keyboard was allegedly produced in limited numbers by Fred Sholes in 1923 but not apparently marketed.

DISSYLLABLE

A shorthand machine with twelve keys produced by a Swiss called Persiaux in 1920. Up to five keys could be used simultaneously at speeds claimed to be up to 180 words a minute.

DOLLAR

A primitive type wheel device subject of several patents from 1890 onwards and manufactured by Robert Ingersoll of New York.

Few have survived, so the price may reach well into three figures or so. **(159 [lower l])**

DWF

A thrust action machine with three row keyboard manufactured briefly from 1923 by Berlin-Karlsruher Industrie-Werke A. G., Berlin, previously Deutschen Waffen-und Munitions-Fabriken, whose initials gave the machine its name.

Not an exciting machine but rare and no doubt interesting to specialist collectors. Price is likely to be around the four figure mark.

EAGLE

An inexpensive swinging sector machine with a three row keyboard and double shift patented by Charles J. Paulson from 1902 onwards and manufactured by Eagle Typewriter Co. of New York. Also marketed as the Defi and the Secretary and later, in slightly modified form, as the Sterling **(250)**.

$2,000 or so.

ECLIPSE

A three row keyboard down stroke, with type bars behind the platen, virtually identical to the Brooks.

Very rare. Five figures.

EDELMANN

This fine example of a type wheel machine with indicator was a German instrument produced first by Wernicke, Edelmann & Co. in 1897, later by others, and still being sold as late as 1914. A small knob on the end of the indicator arm at the front of the machine selected the letter and was pressed down for printing. Inking was by roller. Considerable numbers were produced. An example previously in the author's collection was numbered 16, 549 and other surviving examples have numbers considerably higher than that. Also marketed under the name Gladstone **(164)**.

$1,500 ought to do it.

EDISON MIMEOGRAPH

An unusual up stroke type plunger machine invented by Thomas Edison, the Mimeograph consisted of a horizontal wheel with the type plungers located vertically around its perimeter. This was fitted to a larger wheel in the base of the instrument which was turned to select upper or lower case letters, figures, and characters from an index on the front. Typing was non-visible and was revealed by lifting the carriage, and printing was performed by pressing a key which raised a hammer and struck the type plunger up against the paper, knocking the plunger back into place as the hammer fell. Contrary to some reports, the type plungers were not spring loaded, nor did they return to their original positions by gravity or of their own accord, but only after they were struck by the release of the printing key. Ribbon was used for the impression.

A. B. Dick of Chicago manufactured the machine from 1894, the patents dated the following year, but it was something of an anachronism at that late date and was not one of Edison's most inspired contributions to the history of technology. Despite this, it was manufactured in limited numbers through three model changes of 78, 86, and 90 plungers respectively. The manufacturer's original intention had been to offer the machine as a companion to their Mimeograph duplicating device which they were selling in large numbers—the hammer blows on the type being specifically designed for cutting stencils—but pressure from rival typewriter manufacturers threatening to boycott the company's duplicator speeded the machine's demise.

Certainly, the competition took a dim view of Edison's knuckling in on their territory. Oden's (Underwood-orientated) account suggests that the Mimeograph 'did not possess quality, and although cheap, it was expensive for the owner because it did no better work than other machines, and was not equal to the requirements of general office work. Almost all machines of the day in which it was manufactured would cut a stencil quite as good as the Edison and as a result it was soon discontinued. The fact that they tried to enter the regular typewriter field served to encourage mimeograph competition by typewriter companies who were building machines that would cut satisfactory stencil copies.'[72]

Don't be shocked by prices up to $8,000 or so. **(17, 151, 152)**

EDLAND

A machine with type on the ends of flexible segments of a disk which turned beneath a circular index was patented in 1891 by Joe Edland and manufactured by the Liberty Manufacturing Company of New York. An indicator arm selected the desired character and pressing it down performed the printing. The machine was rough and the castings of poor quality; few were made, few (if any) were sold, and fewer still have survived. In fact, those which have, including an example previously in the author's collection, are all from a small hoard of unsold stock which was discovered some decades ago in the United States.

A later type wheel model with a semi circular index was introduced in 1894 but did little to improve the manufacturer's fortunes.

Too rare to quote. An example previously in the author's collection is only one of very few known to have survived, and none has as yet been offered on the open market. **(153 a & b)**

178. Ideal Self-Tuition Keyboard

ELECTROMATIC

Groundwork on this forerunner of the IBM dates back to 1923 and the involvement of Russell G. Thompson and the Northeast Electric Company in perfecting an electrically operated typewriter which James F. Smathers of Kansas had been working on for almost a decade. Electromatic Typewriters Inc. made little commercial progress, however, until they were taken over by International Business Machines in 1933 and the first commercially successful electrics were marketed soon afterwards. The machine retained the use of type bars, although refinements such as differential spacing were progressively introduced, until the advent of the 'Golf Ball' model in the early '60s for which IBM, in a euphoric rush of blood, made some extravagant claims of paternity concealing or ignoring the fact that the individual elements of the design had all been invented way back in the 19th century.

A couple of hundred should be ample.

ELLIOT-FISHER

The two principal manufacturers of typewriters designed to print on the bound pages of books or on flat sheets of paper were the Elliot Hatch Book Typewriter Co. of New York and the Fisher Book Typewriter Co. of Cleveland, Ohio, and they merged in 1903 to form the Elliot Fisher Co. The machine they produced was a down stroke with a four row keyboard which travelled on rails from left to right as typing progressed. Minor model changes were introduced and sophistications such as tabulators added but the principles remained the same. Some were offered on their own heavy cast iron tables, with a central pedal for lowering the level of the table top for paper insertion. **(155)**

Expect prices up to $1,000.

ELLIS

A combined adding machine and typewriter of imposing dimensions was invented by Halcolm *(sic)* Ellis who was later to design the French MAP machine. It was manufactured from 1910 by the Ellis Adding Typewriter Co. of Newark, New Jersey. It had a few separate keys which typed whole words such as Credit and Debit.

A few hundred should do it.

EMERSON

A front stroke machine with a three row keyboard and a peculiar type bar action was invented in 1907 by Richard Uhlig and produced by the Emerson Typewriter Co., first in Boston and later in Chicago. The unusual feature of the machine was that the type bars stood vertically to either side of the printing point and swung horizontally on their pivots when the corresponding key was pressed. The manufacturer sold out in 1910 to Sears Roebuck who first formed the Roebuck Typewriter Co. to handle the machine, and then four years later, ownership passed to the Woodstock Typewriter Co. which dropped the Emerson in favour of the front stroke upright to which it gave its name. The existing stock in trade of the Emerson operation was bought up by a character called Harry A. Smith who made something of a habit of buying up moribund typewriter concerns and who sold off remaining Emersons under the name Smith.

$500-800 should be enough to buy a good one. **(156)**

EMKA

An Albus machine was sold in Germany under this name which was derived from the initials of the agent, Max Keller.

Cheapie. $300 or so.

EMPIRE

A truly great machine in every respect but such is the perversity of typewriter history that its very greatness has been the cause of its humble end. But the time has come for the Empire to strike back!

Wellington Parker Kidder was responsible for this thrust action machine with its three row keyboard and double shift which he patented in 1892. He did not actually invent the thrust principle itself—the action whereby the type was pressed rather than struck against the platen—for the Rapid holds this honour. But he can certainly claim credit for its successful commercial exploitation.

On the Empire, type bars were located horizontally and fanned out in a sector of which the apex was the printing point. Three row keyboard with double shift, ribbon inking. The shift key raised the platen one or two levels as selected, the type bars remaining on the same horizontal plane. Williams Manufacturing Company of Montreal was responsible for marketing the machine as the Empire, and the same company in Plattsburg, New York, sold it as the Wellington. There were several models, including the larger and more enclosed Model 2 in 1908 and the smaller portable in 1916.

It was an excellently designed and solidly made machine and sold in its many hundreds of thousands all of which (or so it sometimes appears!) have somehow contrived to survive. This alone is responsible for its humble place somewhere near the bottom of the scale of desirability as a collector's item...down there somewhere together with the Blicks and the Rem 7s and the Folding Coronas.

It was manufactured under licence with spectacular success in Germany as the Adler and was also retailed in various parts of the world under the names British Empire, Davis, Lindeteves, and Wanamaker.

$200-300 or so should buy an exceedingly good one, otherwise think in terms of $100-150. **(85, 157)**

ENGLISH

An interesting down stroke instrument with a curved two row keyboard and double shift was patented in 1890 by Michael Hern and Morgan Donne and produced by the English Typewriter Co., Ltd. The interesting feature of the design was the fact that the type bars were maintained in the upright position by means of a counterweight below their pivots and typing was performed when they were struck by the key lever whereupon they returned to their initial position by gravity alone. The machine was not a success and an improved model announced shortly before its demise never made it to the mar-

179. Imperial **(1)**

ketplace. As a result 'specimens of the English are to be met with very cheaply,' wrote Geo Mares in 1909.[62] Not any more! Too rare to quote, but surely five figures.

ERIKA
A three row front stroke folding typewriter manufactured from 1910 to 1927 by Seidel & Naumann A. G. after which it was replaced by a conventional four row model. This German version of the folding Corona was also labelled Bijou, Bijou Folding, and Gloria and was produced in considerable numbers. A 1913 company leaflet claims annual production figures of 100,000 sewing machines, 20,000 bicycles, and 20,000 typewriters, adding machines, and calculators from 3,000 workers.
$150?

ESSEX
The Essex Universal Typewriter Co. of New York launched this swinging sector machine in 1890, but with limited success. It had a two row keyboard with sixteen keys and roller inking, later three rows with twenty-seven keys, double shift and ribbon, the spools tucked neatly away to the bottom right of the machine. The type sector was off-set to the right hand side making for better visibility of the printing. The first model used a hammer to strike type against paper from inside the type sector, but on later models the carriage was made to rock forward for printing.
Too rare to quote, and therefore comfortably into five figures.

EUREKA
The German sewing machine manufacturer Karl Heinrich Kochendörfer's first venture into typewriter production was an instrument clearly inspired by the Hall but probably some ten or fifteen years too late for any serious chance of commercial success. Introduced in 1898, its only significant difference from the Hall was the substitution of an elliptical index to replace the square one, as well as a simplified escapement based on a rack and pawl. Other than that, the type was rubber as on the Hall, characters were selected and printed by means of a small vertical indicator and typing was on the flat page beneath.
Five figures, for sure.

EVEREST
An Italian upright with four row keyboard of conventional design made from the early 1930s by the Milanese company which was already producing the Sabb (previously Juventa). There were numerous models with production continuing until the 1960s when Olivetti, which had taken it over, phased it out.
Cheapie. $80 is probably too much.

FAKTOTUM
The first model by this name was essentially a copy of the down stroke Imperial Model A made in Germany by Fabig & Barschel in 1912. Like the Imperial, the original Faktotum had a curved three row keyboard; a second model introduced the following year substituted a straight three row. It was also labelled Leframa and Forte-Type.
Up to $1,000.

FAMOS
A primitive circular index machine with a vertical type wheel turned by a knob, a single key bringing type into contact with paper and printing around the platen, rather than along its length. It was roller inked, and manufactured from 1910 to

180. International (Courtesy of Bernard Williams)

1913 in Germany by Gustav Tietze A. G. Also marketed in France under the name Victoria.
A totally different machine, better known as Geniatus, was also later marketed under the name Famos.
Flimsy and cheaply made but relatively rare, so some have been known to change hands at prices well into the hundreds of dollars. **(268 [I])**

FAY-SHO
Fay-Sho, and the more complete form Fay-Sholes, were the names under which the Rem-Sho was briefly marketed when early judgements in the on-going legal battles between Remington Standard interests (which no longer belonged to the Remington family) and the Remington-Sholes Typewriter Co. (which belonged to Remington's son Franklin and Sholes' son Zalmon) were decided in favour of the Remington Standard. At issue was the use of the word Remington in the name of a typewriter. Charles Fay, as president of the company, used his own name and the company was called Fay-Sholes Typewriter Co. until appeals all the way to the Supreme Court resulted in a final decision against Remington Standard interests, whereupon the names reverted to their previous form and the machine was once again called the Rem-Sho.
Condition of the machine would determine a precise value somewhere around $1,000.

FEDERAL
A conventional upright which appeared in 1919 as the new model of the Visigraph after the Federal Adding Machine Co. of New York bought up the assets of the former manufacturer. It had only limited production.
Rare but not interesting enough to be worth more than high two or low three figures.

FELIO
A conventional front stroke machine with four row keyboard of low profile similar to that of the early flat bed Royal. The design began its life in Germany as the Nora of 1914 which, like its original manufacturer, became a casualty of the First World War. In 1919 it moved to Amsterdam after the assets of the German concern were bought up by the company which gave the machine its new name. It was also marketed under the name Berolina, but production continued only for a limited time.
Cheapie. Low three figures is plenty.

FIDAT
An Italian upright developed from the Minerva and introduced briefly in 1914 by a Turin manufacturer. The Juventa was also marketed under this name.
Very little value, except perhaps to an Italian collector.

FISHER
Robert J. Fisher patented this down stroke book machine in 1896. Manufactured by the Fisher Book Typewriter Co. of Cleveland, the first model was a cumbersome affair with double keyboard which rode on rails over the flat surface below, printing by down stroke. Later models abandoned the double keyboard in favour of a standard four row universal and the machine was redesigned to present a more graceful appearance. A cylindrical platen mounted in its own sub frame was offered as an optional extra, thus converting the machine to more conventional use, and it is in fact listed in F. S. Webster Company's 1898 ribbon catalogue as 'The Fisher Book and Letter Typewriter.'
The company merged in 1903 with its competitors Elliot-Hatch Book Typewriter Co. to market a single machine called Elliot-Fisher.

181. Junior (Courtesy of This Olde Office, Cathedral City, California)

Early models are very rare and one would expect them to change hands for four figure sums; later models for three figures.

FITCH

This desirable machine with its unmistakable profile was patented by Eugene Fitch of Des Moines in 1886 and produced from 1891 by the Brady Manufacturing Co. of Brooklyn. The Fitch Type Writer, as it was called even at that late date, was of down stroke design with the type bars located obliquely above and behind the platen. This arrangement, in common with that of other comparable designs, made it necessary for the paper to be rolled into open frame baskets on either side of the platen.

First machines had a curved two row keyboard with double shift, later replaced by a straight three row with its own peculiar letter order. It was inked by roller, against which the type brushed on its way to the paper. This clumsy arrangement may well have contributed to the machine's early disappearance although it lasted a little longer on the other side of the Atlantic where examples bore the label The Fitch Type Writer Company, England.

Think $10,000-15,000 or so to secure one. **(78, 160)**

FONTANA

S. A. Fratelli Fontana manufactured this conventional upright in Turin briefly from 1921 until it became the Hesperia the following year. The Fontana was itself a development of the pre-war Fidat.

Cheapie. Even less than the Fidat. A handful of thousands should be ample...Italian *lire*, that is...

FORD

A thrust action machine with a difference, the type bars being pivoted and describing an arc on their way to the platen with a guide raising or lowering them for change of case. With a curved three row keyboard, and the mechanism concealed behind its characteristic decorative housing, the Ford was patented by E. A. Ford of New York in 1892 and manufactured by the Ford Typewriter Co. in 1895. A portable aluminum model was also made. It was assembled and marketed in France under the name Hurtu and in Germany as the Knoch. You will probably need to go to at least $5,000 or more for this interesting machine, although occasionally inflated prices as much as two to three times higher have been reported.

FORTUNA

A conventional upright with four row keyboard, produced in Germany by the munitions manufacturer J. P. Sauer & Sohn of Suhl. It was originally marketed in 1923 as the Stolzenberg-Fortuna after the office equipment company Stolzenberg which had sole retail rights, later sold simply under the name Fortuna. Marketed in England and France as Oliver, the cross-pollination resulting from the fact that Stolzenberg were Oliver's German distributors.

$50-100 maybe, on a sunny day.

FOX

William R. Fox and Glenn J. Barrett designed this blind up stroke machine for which they were granted a patent in 1898 with manufacture by the Fox Typewriter Co. of Grand Rapids Michigan, beginning the same year. It was a well made instrument, similar to other up strokes of its time except that change of case was performed by the lateral movement only of the platen and not of the entire carriage. It was subjected to only minor changes until the 1906 Model 24 when the machine was redesigned as a front stroke visible, production continuing until 1921. A Fox 23 labelled Rapid No. 10 is recorded.

William Fox retired from the company in 1915 and this was reorganized under new management. The name was retained, however, and a folding portable model appropriately labelled Fox Portable and Baby Fox was introduced in 1917. This machine featured a carriage which hinged back behind the body giving it a flat profile, but it was relatively short-lived due to legal action claiming infringement by the Corona Typewriter Co. which was producing its famous folding portable in vast numbers, even though the folding feature of the two machines was totally different.

A conventional portable with three row keyboard and double shift was substituted in 1920 and named Fox Sterling. This model no longer folded, but the manufacturer did—the following year.

Many different models. Earliest might even make up to four figures, in a brisk breeze, but later models get progressively cheaper, down to low hundreds. Folding portables fetch in the region of $300 or so. **(10 [l], 161, 162)**

FRANCONIA

A conventional German upright designed in 1909 by Carl Fr Kührt of Nürnberg and manufactured from 1911 by the firm

182. Karli (Courtesy of Dresden Technical University)

of Otto Baldamus of Koburg until the First World War. It was also marketed under the name Excelsior. From 1919 it was made by Mayer & Co. of Augsburg under the name Omega. Very little value. Low three figures should buy you a good one.

FRANKLIN

A fine down stroke machine in which the ends of the type bars and the key levers meshed with each other, as opposed to being attached in the conventional manner. It was invented by Wellington Parker Kidder and covered by a patent filed in 1889, granted two years later and assigned to the Tilton Manufacturing Co. of Boston. It was marketed by the Franklin Typewriter Co. while Kidder himself, shortly thereafter, went on to invent the great machine to which he gave his Christian name. The Franklin originally had a two row and later a three row keyboard. Inking was by ribbon, with the spools just a little too intrusive. Later examples were labelled New Franklin. A model with a straight conventional keyboard was being developed in 1906 but the following year, before it could be introduced, the company was bought up by the Victor Typewriter Co. of New York and the make was phased out.

Expect to pay in the $800+ range, but good early ones may fetch well over $1,000. **(83)**

FRISTER & ROSSMANN

The Berlin sewing machine manufacturer Frister & Rossmann produced this German version of the up stroke Caligraph under licence from the American manufacturers from 1892 to 1905.

$1,000-$1,500, but you may find the corresponding Caligraph equivalent a better value.

GALESBURG

A third name for the type sleeve machine which began as the Munson and then became the Chicago until 1912 when manufacture was taken over by the Galesburg Writing Machine Co. of Galesburg Illinois. The design changed remarkably little during the first twenty-five years of manufacture from the 1890 appearance of the first Munson, but 1915 saw the introduction of an aluminum model, with manufacture finally closing down altogether in 1917.

Prices as for Chicago.

GARBELL

A small thrust action machine with three row keyboard, manufactured from 1919-1923 by Garbell Typewriter Corporation of Chicago. Relatively rare, so probably worth around the four figure mark.

GARDNER

An attempt at simplifying typewriter operation and manufacture which went wide of its mark was proposed by John Gardner of Manchester with a design for which he was eventually granted a patent in 1899, by which time the limited production of the instrument had already ended in failure. In attempting to economise on parts and cost, the Gardner used a mere fourteen keys and a space bar. Each key served for two letters, one of which was on the key-top in black for the selection of which the simple pressing of the key was required, while the other was in red and was selected by simultaneously pressing the key and the space bar. It had a vertical type sleeve and printing was by means of a hammer striking the paper against it from the rear. Inking was by roller, and shift selection by means of a rod located beneath the top plate. Manufacture lasted a few years and ended altogether in 1895 when the company went into liquidation, but attempts at promoting sales on the Continent continued a while longer. Sold in France under the name Victorieuse and in Germany as the Victoria. If you buy one for less than five figures, you have done the deal of your life! **(70)**

GENIATUS

This is a machine which on face value had no justification for even being manufactured at all, given its late date, and yet it enjoyed considerable success for the short time it was marketed. It was a small lightweight affair, with swinging sector, indicator, curved letter scale, double shift, and ribbon inking. The type was on a vulcanized rubber strip divided vertically into segments with three characters per segment. This type strip was passed over a roller at the printing point and anchored at both ends of the swinging sector. Printing was by means of depression of the selector key.

Interestingly enough, the manufacturer's name does not appear on the machine and yet all examples were meticulously stamped with production dates, and with serial numbers running well into five figures. It is believed to have been manufactured by Gundka-Werk GmbH and was also marketed under the names Geka, Famos, and Gloria. From two examples previously in the author's collection—serial no. 5, 612 dated 25/5/28 and no. 9, 344 dated 21/8/28—a production of over 1,000 units per month is evident.

Anything up to $300-400 is about right. **(163)**

GERDA

A German machine manufactured in 1919 by Georg Emig of Berlin whose pre-war connection with Blickensderfer no doubt facilitated his use of their type wheel on the machine he designed specifically for blind and wounded war veterans. The typing hand rests on a swinging plate geared to the type wheel with a curved letter index at the top and an aperture through which the index finger could feel the braille cell corresponding to the letter selected. The whole plate was pressed down for printing, with inking by roller.

Relatively rare, so don't be shocked by prices in the low four figures.

183. Keystone (Courtesy of Bernard Williams)

GERMANIA

A Jewett typewriter assembled in Germany from components imported from the US from 1898 onwards. It was also labelled Germania-Jewett.

$800, perhaps?

GISELA

Gisela Schreibmaschinenwerk Günter & Co. manufactured a three row front stroke portable in Germany in 1921.

Cheapie, which means a high two or low three figure price tag.

GLASSHÜTTE

One of the world's great centres of watchmaking and precision engineering also produced a typewriter for a few years after 1922 when the Schreibmaschinenindustrie Glasshütte GmbH launched a conventional four row upright. It was also marketed as the Usapax-Visityp. The exercise was undertaken to use up excess industrial potential in the area, but was not commercially viable in the long term.

An oak box 13 x 5 inches with a 'feeder' served to contain the paper.

'The human hand will perform this operation (i.e. typing) far better than any set of wires ever invented by man,' said the blurb. '…noise is avoided, simplicity obtained, and novelty secured' with 'even greater rapidity than is attained in the most approved machines…For example a sentence such as 'the action of offering land to the king…' (presumably, just the sort of phrase the average man might need every day of the week!) '…is written in 16 instead of 39 strokes.'

Rare, but so what? Novelty value, perhaps. If you already have everything else or have unlimited finance, what you pay depends on your degree of insanity.

GOURLAND

A conventional front stroke portable produced from 1920 by the Gourland Typewriter Corporation of New York. Charles Spiro of Columbia type wheel fame was responsible for the design and with that impeccable pedigree one might have ex-

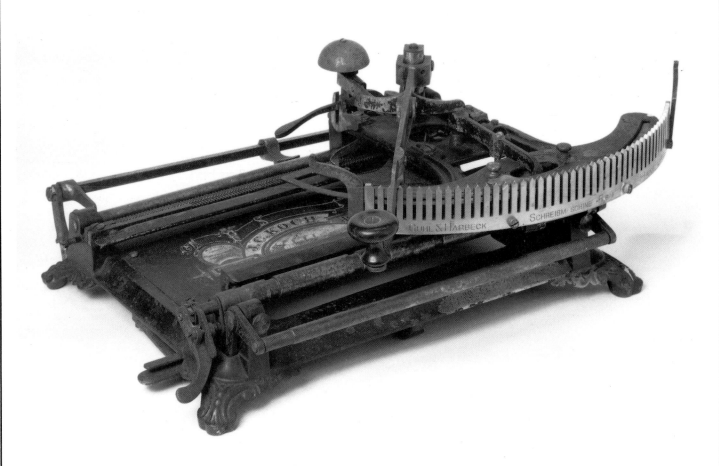

184. Kosmopolit (Courtesy of Phillips, London)

A couple of hundred, perhaps? Glasshütte is a magic name to many people, horological enthusiasts for example, so anything from there is collectable.

GLOVE

Hardly mechanical although it was promoted as a typewriter in an 1891 advertisement, the Glove or Glovegraph invented by a Mr. Cary consisted merely of a leather glove with rubber type attached down the fingers, following (quite inexplicably) the by then familiar QWERTY order. Lower case on the inside of the glove and, would you believe it, upper case on the back with 'AND,' 'THE,' 'OF,' 'ING,' and 'TION' down the thumb!

pected the machine to prove a better seller than it was. Some examples bear the label Alexander, also Wright Speedwriter. Somewhere between $100-200 ought to be sufficient.

GRANVILLE AUTOMATIC

So called by its American inventor Bernard Granville because all functions of the machine except insertion and removal of paper could be performed 'automatically' from the keyboard, this thrust action design with standard four row keyboard was patented from 1891 and manufactured five years later first by the Granville Manufacturing Co. and then by Mossberg and Granville Manufacturing Co., both of Providence, Rhode Is-

land. The thrust action principle, later immortalized in such machines as Empire, Adler, Noiseless, etc. was actually Granville's original invention, patented in 1888 and used on his Rapid.

Rare machine. Anything up to $5,000 or so is OK.

GRAPHIC

One of several German machines virtually identical in principle if not in appearance to the Hall Type-Writer, the Graphic was made in 1895 by C F Kindermann & Co., Berlin, later by Heinr. Bonnin. It was a heavier machine than the Hall and more decorative, the letter index was curved and not square and the individual letters were more easily legible than on the American machine whose letters could be read only if viewed directly from above.

A second model offering some improvement in typing quality was later introduced.

Probably well over $5,000...well over... **(166)**

GRÖNBERG

A single reference to a German machine by this name states that it was similar to the Soennecken and the Brackelsberg and a patent drawing of the former indicates only that the machine appears to have been a linear index device.[12] Assuming the obvious, namely that the reference is to Brackelberg's Westphalia and not to the multi-keyboard syllable device to which he gave his name, one might infer that the Grönberg was indeed a linear index or linear type plunger machine, apparently similar to the Merritt, of blind up stroke operation.

Price would be well into five figures...if an example were ever to surface, but we have all said that about many machines in the past and they have (surfaced, that is), so presumably there may well be one out there somewhere.

GUNDKA

Gundka-Werk GmbH mass produced this simple type wheel machine invented in 1924 by Paul Muchajer which, despite appearances, was not a toy but was intended for serious work. An indicator operated by the right hand selected the desired letter from a curved index; pressing down on the indicator knob brought the type wheel into contact with paper, the slots in the front of the machine serving as guides. It had ribbon inking, with keys on the left for spacing and change of case.

The type wheel and letter index were easily interchangeable and the machine, under a variety of names, was widely sold throughout the world. Minor changes in appearance and quality are apparent between early and later (post 1926) models which have confusing designations, model numbers and names under which the instrument was sold in different countries. Frolio models 5 and 7 are known, for instance, but not the earlier numbers nor the intermediate one.

Marketed under the following names: Bambino, Frolio, Gefro, G & K, Gundka, MW, Perlita, Scripta and Write Easy. Machines in general are notoriously bad at respecting international borders, but from observing market patterns it appears to the author that Frolio was probably intended for the Italian market, Perlita for Spain, Write Easy for England, and G & K and Gundka for the German home market. Frolios were to be found in considerable numbers in Italy some twenty-five or thirty years ago and in fact a Frolio 7 from the Rome flea market at Porta Portese back in the '60s was the first machine the author ever bought. Price: 100 lire.

A cheaply made copy of the machine labelled simply Junior was made in England by EMG of Dover, apparently during the 1930s or so, and marketed as a toy.

Common but cute. Anything up to $200-300, perhaps. **(6 [r], 277)**

GYNEE CIPHO

The Gynee Cipho Typewriter Co. of London marketed this primitive vertical type wheel machine which retailed, predictably enough, for one guinea.

Prices would be in the Simplex bracket, if any have survived.

HAGELIN

This most sophisticated of all strictly mechanical printing cipher machines was developed by a Swede called Boris Caesar Wilhelm Hagelin after the First World War. Six cipher wheels on the front of the machine, rotated to the desired cipher, turn a cage of twenty-five (later twenty-seven) bars with moveable lugs on each bar. A knob turned by the left hand selects the plaintext letter on a wheel visible through a window in the cover; a handle on the right turns the cage which turns the cipher wheels to select the required cipher letter on a small type wheel, which then prints it on paper tape. Roller inking. Different models were produced and some, such as the

Cryptos, printed both cipher and plaintext on separate tapes. Essentially, codes can most easily be cracked when the encoding sequence repeats itself: on the Cryptos, this does not happen until after 100 million successive letters have been enciphered! Quite rare but probably not particularly popular. Somewhere in the region of the mid-three figures ought to be enough. **(167)**

HALDA

Halda Frickurfabriks A. B. of Sweden produced very limited numbers of an up stroke machine, with four row keyboard patterned after the Densmore, from 1896 onwards. It went through several model changes until the successful Model 4 in 1902 launched the make with increased sales and production. It was not until the Model 8 was introduced in 1914 that the machine became a conventional visible upright.

Early model is rare and correspondingly priced. Later ones are of little value.

HALL BRAILLE WRITER

A fine, sturdy, compact and well-made typewriter for the blind, with three piano-type keys per hand plus central space key, for embossing braille cells. Inventors were Frank H. Hall, G. A. Sieber, and T. B. Harrison and the date was 1891.

Excellent value at the current price of $300-500. **(168)**

HALL TYPE-WRITER

The first commercially successful index machine was patented by Thomas Hall of Brooklyn in 1881. Hall's involvement with the development of the typewriter began back in the 1850s and first resulted in a down stroke type bar machine with circular keyboard which he patented in 1867 but which he never quite succeeded in getting off the ground.

The Hall Type-Writer Co. of Salem, Massachusetts, however, began manufacturing the 1881 index device in the same year in which it was patented and thousands were sold up to as late as the turn of the century, undergoing several minor changes in design and spawning a host of thinly disguised foreign copies. Of Halls alone, *Scientific American* in 1887 quotes production figures of 300 up to 1883, 3,000 to 1884, and 5,000 to 1886.[83]

The machine used a square rubber index attached to an indicator arm. Selection of the desired letter brought the rubber type directly over the printing point, a mask ensuring that the remaining letters were blanked off. Inking was by pad and typing was achieved by pushing down the handle and thus the whole top plate, with a very small block the size of the individual letters pressing the rubber type onto the page. This whole assembly travelled from left to right as typing progressed, along a rod which meshed with a wheel above the index, while the platen and paper remained stationary.

The machine was fitted into an elegant wooden case and could be tilted forward to improve visibility since it was virtually impossible for the typist to read the letters she was selecting unless her face was directly above them. A separate and more easily legible index screwed onto the machine beside the original one, with a clumsy indicator extension arm for letter selection, was only one of several later 'improvements' made to the machine in attempts at prolonging its useful life span and an even more radical modification, dated 1889 and labelled New Model, offered metal type bonded to the rubber plate and a carriage which moved while the index remained stationary.

Not a rare machine but an interesting one, and reasonably priced at up to four figures. **(5, 72)**

HAMMOND

This great machine by one of the pioneers of typewriter history was first patented in 1880 after more than a decade of developmental work. James Bartlett Hammond was the man responsible and the Hammond Typewriter Co. of New York began producing his machines from 1881.

The design was a radical departure from the type bar principles which his contemporaries at Remington were busily promoting. Hammond, recognising the limitations inherent in type bar design (poor alignment, irregularity of impression, slowness of operation, clashing of bars, and so on), fitted his type to a swinging sector which moved through only a relatively small arc to locate the desired character at the printing point whereupon a hammer situated at the back of the machine

187. Merkur (Courtesy of Arthur Frehner)

188. Merritt

189. Mignon 2

was released to strike the paper against the type. The action of the hammer was completely independent of the pressure applied to the keyboard, thereby ensuring perfect regularity of print. Inking was by ribbon.

Built into the design, however, was its only serious flaw. With a type sector in front of the page and the hammer behind it, there was no alternative but to roll a fresh sheet of paper into a coil, insert it into a cylindrical basket and pass it up through the machine between feed rollers for typing. The design retained this feature throughout its long and illustrious career and it is further testimony to the excellence of the machine that this one serious disadvantage was tolerated.

Early models had curved two row keyboards, first with thick piano-type keys, later with smaller rectangular keys and finally with conventional ones, by which time Hammond had bowed to the competition and was also offering models with optional straight three row keyboards. The letter order was originally the DHIATENSOR 'Ideal' arrangement and was followed by a period during which the QWERTY 'Universal' was offered as an option before industry standardization made the former obsolete.

Capitalizing on the ease with which different type shuttles could be interchanged on the machine, Hammond offered literally hundreds of alternatives covering languages, alphabets, scripts, scientific symbols, ciphers and so on, making the machine the most versatile on the market. Models with dual Oriental and conventional keyboards had a special device permitting the carriage to be moved from left to right at the will of the operator, others with triple shift printed up to 120 different characters, mathematical models with special shift and half back space facilities permitted the typing of such things as numerators, denominators, square roots, and nth powers, and so on. Only the Blickensderfer and Mignon offered comparable if less sophisticated features.

Hammond type shuttles themselves underwent several modifications before standardization was achieved. The first design consisted of two separate sectors on a central spindle. This was followed by a single shuttle fitted into a slot in an incomplete wheel, and finally by two diametrically opposed shuttles (eg. one standard and one perhaps italic or mathematical) fitted to a complete wheel and selected by lifting the spindle and turning through 180°.

If this sounds confusing, then attempting to place all the modifications and models into their correct chronological sequence

has proved even more difficult. The piano keyboard model with laterally enclosed keyboard, the levers partially covered by a curved flat cover, was certainly the first and serial numbers extend into the 20,000s. The cover was then dropped from all future models but the piano keyboard retained until the 1893 Model 2 with curved or straight keyboard, followed it seems, by the Model 12 still with optional curved or straight keyboards. A 'New Hammond No 12' was announced in 1907, Multiplex in 1915, Folding portable in 1923, until the machine ultimately became the VariTyper in 1927, the company changing its name to VariTyper Inc. from 1931 and later to Ralph C. Coxhead Corp of New York.[91]

An electric Hammond was also briefly introduced at a time when most major manufacturers were jumping on this bandwagon. All of these units consisted essentially of a small electric motor attached to an otherwise standard machine, the motor serving only for carriage return.

Vast spread of models, with spread of prices all the way from $2,000 or so down to $300. **(50, 64, 71, 74, 81, 169, 170, 171)**

HAMMONIA

The first typewriter to be mass produced in Europe was a remarkable linear index device in which type was cast in a row and attached to the underside of a blade with a handle on top and an indicator on one side. The blade ran in a slot at right angles to the line it was printing until the indicator pointed to the desired letter, whereupon the handle was pressed down for contact between type and paper, a rack sliding the blade a space to the right. It was produced in 1884 by the German sewing machine manufacturers Guhl & Harbeck of Hamburg who mounted it on a base of obvious sewing machine inspiration. The first patent covering the design is British and dated 1882, in the name of Andrew Hansen. A German patent in the name of H. A. Guhl of Hamburg is dated the following year.

Pay the asking price, however high it is, and run for it! If the price is less than $15,000-20,000, you have done OK. **(18, 172)**

HARRIS

A conventional front stroke machine also called Harris Visible with three row keyboard and double shift, invented by D. C. Harris in 1911 and produced in the US by the Harris Typewriter Company. Sears Roebuck sold it by mail order and the American Can Co. also handled distribution. In 1914 the manu-

190. Mignon 2
(Courtesy of
Phillips,
London)

191. Molle 3
(Courtesy
of Bernard
Williams)

facturer changed its name to Rex Typewriter Co. and the machine was marketed under the names Autocrat, Reporters Special, Rex, and Rex Visible.

No more than a few hundred. **(226)**

HARRY A. SMITH

A conventional front stroke upright originally known as Bar-Blick until the demise of the Blickensderfer enterprise in 1917, whereupon the production of the machine was taken over by Harry A Smith, who made a career of buying up moribund typewriter manufacturers, and who marketed the product under his own name.

$100 or so?

HARTFORD

John M. Fairfield of Hartford, Connecticut, designed this up stroke machine with double keyboard for which he was granted a patent in 1895. The inventor was formerly with the Caligraph factory and his design was clearly influenced by that make. 'Blind' machines printed underneath the platen which had to be raised to make the typing legible: in order to facilitate this operation on the Hartford, the platen was spring loaded so that it swung up automatically when a lever was pressed although it had then to be repositioned manually. Type was curved to match the profile of the platen. The original price was the 'mandatory' $100 for machines of that design, a price which the manufacturer soon halved, thereby incurring a barrage of abuse from his incensed competitors. Model 2 appeared, with minor modifications, in 1896. The four row keyboard Model 3, also marketed under the name Cleveland after production moved to that Ohio town, appeared in 1905. $2,000-2,500 ought to be plenty.

HASSIA

Forerunner of the Torpedo, this conventional German upright manufactured from 1904 by Jean Voelker & Co. of Neu-Isenburg featured a hinged carriage permitting easy access to the innards.

Loose change.

HELIOS

A cylindrical type wheel of relatively small diameter with characters in four rows was at the heart of this compact German machine with two row keyboard and triple shift first manufactured in 1908 by Justin Bamberger and the Deutsches Schreibmaschinenwerke, Munich (the company previously producing a circular index machine called Liliput). It was next made by Helios Schreibmaschinen GmbH, a subsidiary of Kanzler Schreibmaschinen A. G. (which was making the Kanzler) in 1909, and finally, five years later, by A. Ney of Berlin who called it Helios Klimax. Plans for the manufacture

192. Monarch Pioneer (Courtesy of Sandy Sellers)

193. Monarch Visible 2 (Courtesy of Sandy Sellers)

194. Morris (Courtesy of Christie's South Kensington, London)

195. Moya (Courtesy of Bernard Williams)

of a three row keyboard model did not materialize. It was also sold under the names Bamberger, Ultima, and Portable Extra. A cute machine, but not overly rare, so $1,200-$1,500 should be enough for a good one. **(7)**

HERMES

These were Swiss conventional front stroke machines with four row keyboards produced by E. Paillard & Co. of St. Croix from 1923 onwards.

HERRINGTON

A primitive vertical type wheel machine with roller inking patented by G. H. Herrington and D. G. Millison in 1884 and mass produced by McClees, Millison & Co. of Chicago. $100-200 ought to be enough.

HESPERIA

A conventional Italian front stroke upright originally named Fontana, marketed for a few years from 1922 by Compagnia Italiana Macchine da Scrivere Hesperia of Torino.
Zzzzzzz.... A two figure price tag is about right...that's dollars, not *lire*.

HOOVEN

An automatic typewriter using perforated paper in rolls comparable to those more commonly associated with player pianos. First placed on the market under the name National Automatic in 1912, then as the Hooven. Manufacturer was Hooven Automatic Typewriter Co., initially of Cincinnati, later of Hamilton, Ohio.
Very, very rare, if indeed any have survived at all.

HORTON

The Horton Typewriter Co. of Toronto was the manufacturer of this oblique front stroke type bar machine with double keyboard and partially elevated carriage, patented in 1883 by journalist and court reporter Edward Elijah Horton. Together with his brother Albert he incorporated the first Canadian typewriter manufacturing company in Toronto in 1885. Production was very limited. It was the first type bar machine on the market to offer *almost* visible typing, even though this was partially obscured by the ribbon (the Daugherty was completely visible).
Very few are known. Five figures, and haggle over how high. **(82, 175)**

HOUSE

This successful printing telegraph was patented in Britain in 1845 by Jacob Brett, and a year later in the US by Royal E. House. The two men are reported to have worked together on the design but the exact circumstances of their co-operation is not known. A piano keyboard controlled a rotating cylinder on the transmitter with a pin on the cylinder corresponding to each key in such a way that pressing a key stopped the cylinder on the transmitter at the selected letter and also stopped a synchronized type wheel in the receiver on the same letter. The keyboard was in alphabetical order and printing was on paper tape.

The design was further developed in later patents but achieved less success in Britain than in the US where House continued promoting it for a long time after the Europeans had given up on it.

None has survived in private hands, as far as the author is aware, so if one surfaces, just pay up.

HUGHES

An important development in the history of the printing telegraph, designed by D. E. Hughes of Kentucky in 1855 and patented a year later. The machine used a piano keyboard built into a table which housed a weight driven clockwork movement to drive a rotating pin barrel. Type wheels on the transmitter and receiver were synchronized so that the same message was recorded at both ends of the line. The concept was not original of course (see House, *inter alia*) but was highly successful and the Hughes was used extensively throughout the world, proving to be the best printing telegraph for the better part of the second half of the 19th century. Froment in France and Siemens in Germany were among those who supplied it.
Very rare, so the price will be four figures for sure. **(45, 46)**

HYNDMAN

An inexpensive index machine sometimes called Hyndman's National and dating from the late 19th century was briefly listed as available on both sides of the Atlantic. It had a limited life, and there are no known survivors.

IBM

See Electromatic.

IDEAL (1)

A fine German oblique front stroke machine with four row keyboard manufactured from 1900 by Seidel & Naumann A. G. of Dresden, although patented by E. E. Barney of Groton, New York. Early examples had the name characteristically cast into the framework on the sides. Carriage return and line spacing were performed in one motion by lowering a handle on the right. Later models were conventional uprights, but special purpose machines such as the Ideal-Polyglott, Ideal-Du-

196. Moya 2

plex, and Ideal-Oriental used a combination of four row keyboard and double shift, making it possible for the machine to accommodate two alphabets.

Very well made machines, so consequently many of the conventional models have survived. Special purpose ones are rare. $400-500 tops, for the early conventional models. **(177)**

IDEAL (2)

An example of a Moya bearing the Ideal name and the logo of Seidel & Naumann has appeared in Germany, probably made under licence from the British company around 1908 or so, after which date the Moya design was made obsolete by the introduction of their Imperial (1).

IDEAL (3)

J. H. Simmons of Croydon produced a wooden dummy of a four row keyboard to teach touch typing without the deafening clatter of a conventional upright typewriter. 'Ideal' for considerate young *typistes*, particularly those concerned about placating irascible landladies. **(178)**

IMPERATOR

A German conventional upright introduced briefly in 1924 by Hegeling-Werke A. G., Eitorf.

IMPERIAL (1)

A down stroke machine with a curved three row keyboard and double shift manufactured from 1908 by the Imperial Typewriter Company, Leicester. The company was formed that same year as a partnership of local business interests and a small local manufacturer who had been making a type sleeve instrument called Moya.

Model A was replaced in 1915 by a Model B which was virtually identical; there was no Model C, but a Model D was announced 'a trifle prematurely' (according to the factory's Catalogue of Historical Typewriters) in 1919 but not produced until 'a full two years later,' i.e. 1921. This model sported a straight three row keyboard and was still a down stroke but with the type bars in a reduced curve—certainly not the most attractive of designs. 1923 saw the introduction of a smaller portable version of the Model D, followed in 1927 by the completely redesigned Model 50 of conventional upright format. Electrics were added to the range in 1960. Model D continued to be made for a short time as Imperial Junior, and a German version of the Model A was sold under the name Faktotum. Ajax, New Imperial, Lloyd, and Typo were also names under which Imperials were sold in different parts of the world.

Good machines but commonly found, hence $200-400 should cover the range. **(94, 179)**

IMPERIAL (2)

Karl Heinrich Kochendörfer of Leipzig, a sewing machine manufacturer already involved in typewriter production with a machine called Eureka, introduced a three row keyboard type wheel instrument with double shift, inking by roller. It was said to have been based on an American design—not

197. Munson (Courtesy of Christie's South Kensington, London)

198. National

surprising, since it bears more than a passing resemblance to the Blickensderfer. Only a few examples are believed to have been manufactured, hence it is likely to fetch four figures.

IMPERIAL (3)

The conventional Remington No 10 was labelled Imperial No 1, allegedly for easier penetration of the British market.

IMPERIAL (4)

A blind up stroke machine with double keyboard introduced briefly in the US in the early 1890s.

IMPERIAL (5)

The Imperial Typewriter Company of Newark was formed in 1904 to re-launch the Manhattan, which was itself a copy of the old up stroke Remington No 2. Each successive reanimation of this design was shorter lived.

Despite the relative scarcity of the name, it is unlikely to make as much as four figures.

IMPROVED TYPE WRITER

Sholes and Glidden Type Writers went through a number of name changes and modifications from the time the first examples fitted on modified Remington sewing machine tables left the factory in 1874 until the design was formalized for years to come as the ubiquitous Remington Model 2 of 1878. Confusion has long surrounded the sequence of the different specifications, compounded by the fact that early machines were often technically up-dated and even aesthetically redecorated and relabelled in later years. Furthermore, models typing upper case only continued to be made and marketed even after the introduction of models offering both upper and lower case.

Improved Type Writer No 1 continued to be marketed as an upper case only machine. Improved Type Writer No 2 was the upper and lower case model; this was then labelled Perfected Type Writer No 2 for a short time and ultimately became the Remington No 2. Improved Type Writer No 4 was an upper case only machine and became the Perfected No. 4.

The simple original title 'Type Writer' was ultimately dropped as the sole name of the machine which the Remington factory was producing, after Crandall, Hammond and others had appeared on the market with their own type writers and Remington was forced to concede that the term had become generic.

The existence of a separate so-called 'open frame' Sholes and Glidden model sometimes referred to in past literature is sus-

200. Noiseless Portable **(l)**, Juventa **(r)** (Courtesy of Bernard Williams)

pect, since the enveloping panels originally fitted to embellish early machines, and coincidentally, to protect their innards from inquisitive fingers, were often later discarded as redundant.

A connoisseur's machine and very, very rare. This is an interim model, and is even rarer than the earlier Sholes and Glidden, although arguably less desirable. Expect to pay in the five figures.

INDEX VISIBLE

A spring-loaded type wheel pulled to the desired character by means of a cord attached to the operator's index finger which selected letters from a fixed 'keyboard' was the essence of this freak introduced in 1901 by the Index Typewriter Co. of New York. It offered ribbon printing, with the carriage brought forward against the type wheel by means of a key on the left. An almost identical machine called New England was marketed in London the previous year.

Exceedingly rare, so in all probability high four figures.

INGERSOLL

Robert Ingersoll of New York, manufacturers of a wide range of inexpensive articles including typewriters such as Simplex and Dollar, also marketed under their own name perhaps the most primitive machine of the lot. This used type on individual blocks riding on a rail and pressed down manually onto a flat sheet of paper on the base, inking by sliding over a pad on the way. It was also sold under the name Nason. 'As a typewriter, it is beneath notice…'[62]

$100-200, perhaps?

INTERNATIONAL (1)

One of several names under which similar machines of linear index design were marketed. The exact relationship between these various brands has not been fully established, although it is inconceivable that rival companies would have been permitted to market with impunity what was essentially an identical concept, given the raging patent wars which were constantly being waged over even the most minor infringements. The inventive genius of Lee S. Burridge in 1885 appears to deserve the credit for the first patent which was marketed as the Sun. Odell arrived on the scene with a comparable design two years later.

The International, marketed by the New American Manufacturing Co. of Chicago and also called New American, appeared in 1890 or so. In common with all the above instruments, it consisted of a linear index traversing a platen of small diameter at 90°. Roller inking. The index housing was pivoted so that the letter selected could be pressed down onto the paper, the same motion advancing the paper by the required amount. $1,500-2,000.

INTERNATIONAL (2)

A totally different and unrelated design to the previous entry but marketed under the same name was an up stroke type bar three row keyboard machine patented by Lucien Stephen Crandall in 1886 and produced three years later. A wide ribbon was used, passing from front to rear rather than from side to side. Two additional models, one with a double keyboard and the other with a four row keyboard, appeared in 1893 but the machine had only limited acceptance and manufacture ceased five years later. It was a heavy and clumsy brute which is the more surprising since Crandall's other invention and the one to which he gave his name was neither of these, although a contemporary publication called *Phonographic World* of New York recorded the view that 'The International and the Crandall had led to the invention by stenographers (i.e. typists) of more swear-words than all the other typewriters combined' and knowing how bad some of them were (the typewriters, that is) makes this judgment the more damning.[62]

Very rare, so don't be shocked by the four figure price. **(180)**

INVICTA

A conventional Italian upright introduced in 1921 by S. A. Invicta of Torino. There were numerous similar models with production continuing until the Second World War. The advertising blurb for the machine was refreshingly and uncharacteristically low key in times when hyperbole was the norm: 'No improvements…no particular features…just a good and well made standard typewriter…'

And boring, from a collector's viewpoint! Loose change ought to buy it, if you must.

JACKSON

Andrew Steiger, one of the men responsible for the design of the Yost, patented this instrument which relied on a similar grasshopper action but unlike the Yost, offered visible typing. It had a four row keyboard and type bars resting on a curved inking pad from which they performed their characteristic leap. Its patent date was 1896 and limited manufacture was begun in 1898 by the Jackson Typewriter Co. of Boston. Another effort to market the design was apparently made in 1903, but without success.

Well into four figures, if you find one. **(77)**

JAPY

A conventional upright manufactured in France by Japy Frères from 1909, after the company had bought up the assets of the Rem-Sho outfit when it had gone into receivership. The Japy was, in fact, a renamed Rem-Sho Model 10.

Ho hum! An investment in the low three figures should get you a beauty.

JEWETT

George Jewett and the Jewett Typewriter Company of Des Moines, Iowa, produced this large double keyboard up stroke machine in 1892. It developed out of the Dennis Duplex with which Jewett was involved and in fact the company was at one time called Duplex-Jewett Typewriter Co., of the same address. There were numerous similar models, the manufacturers persisting with the design until the early years of this century. They even offered intriguing variations on the theme, including what might be called a quadruple keyboard model in 1902, for typing in two alphabets, which sported no fewer than 122 keys. It was also assembled in Germany from 1898 and labelled Germania and Germania-Jewett. A re-modelled Jewett Visible appeared in 1904.

Count on spending $2,000 or more for a good one.

JUNIOR (1)

Charles Bennett of New Jersey patented this intriguing miniature typewriter in 1901 although the instrument was not marketed until 1907. Patents in some European countries had already been granted the previous year. Capable of serious work and designed to be carried comfortably in a brief case or overcoat pocket and not a toy machine in fancy dress, the Junior was of type wheel design with three row keyboard and double shift, roller inking, the type wheel striking against the platen. Touch was irregular as a result of the machine's irritating design which featured overlapping keys, and was further compounded by the fragile hairsprings on which these depended, but the Junior was manufactured in respectable numbers for a couple of years until it was replaced by the slightly improved Bennett. Size was approximately 28 x 13 x 5 cm.

$400-600 should do it. **(181)**

JUNIOR (2)

A German tinplate toy machine, c. 1920, with circular index, type wheel, miniature carriage, and dummy keyboard. The early model had roller inking, later examples used ribbon. Prices up to $200 have been known to change hands for this machine, but you may opt to pitch your own ceiling at…what?…$20? $30? **(268 [r])**

JUNIOR (3)

A cheap copy of the type wheel Gundka labelled simply Junior and apparently dating from the 1930s, or so, was made in England by EMG of Dover and marketed quite obviously as a toy.

$100 or so should be enough. **(6 [l])**

JUVENTA

An Italian portable manufactured by S. A. Industria Dattilografica in Milano in 1922. Three row keyboard, front stroke. It was also marketed as Agar, Agar-Baby, Ardita, Diadema, Fidat, Merkur and Sabb.

Cheapie. **(187, 200 [r])**

KANZLER

An ingenious design and one which might well have resulted in a dainty and delicate machine, given the saving in the number of moving parts, the Kanzler is in fact an instrument of quite daunting proportions if you happen to come across an example in a market and decide to take it home on a crowded

199. New Century 6 (Courtesy of This Olde Office, Cathedral City, California)

201. North's

bus. Invented by Paul Grützman in 1903 and manufactured by A. G. für Schreibmaschinen-Industrie as the Hansa, later by Kanzler Schreibmaschinen A. G., it was of thrust action design with a curved four row keyboard of forty-four keys arranged in eleven radial columns, each one with its own type bar printing a total of eight upper and lower case characters, with case selection by shift key. There were four models offering such minor modifications to specifications as carriage length, until production ended in 1912. Also sold as Kanzler-Rapid and Chancellor.

Expect to part with $800-1,000, or possibly even more, for this magnificent beast. **(3)**

KAPPEL

A conventional German front stroke upright with four row keyboard introduced by Maschinenfabrik Kappel A. G. in 1914. Manufacture continued through many models until the Second World War.

If the asking price is three figures, turn and walk away.

KARLI

A remarkable linear type plunger instrument of which there is only one extant survivor, in the Dresden Technical University. The left hand slides the type carrier to the selected letter and the right presses the typing key. An example was recorded in the Seidel & Naumann museum, which provides the only clue to this company's being the possible manufacturer. It may be one and the same machine as the one in Dresden University.[63]

Too rare to value, but if one comes on the open market you are unlikely to walk away with it for less than a five figure sum. **(182)**

KENBAR

A German portable was sold in England under this name but details are lacking.

KEYSTONE

The Keystone Typewriter Co. of Harrisburg, Pennsylvania, produced this swinging sector instrument patented in 1898. A hammer striking the paper against the type sector from the rear was its salient feature. Three row keyboard, double shift and ribbon. Reported to have been sold in Germany under the name Grundstein.[88]

It will probably set you back in the region of $2,000-3,000. **(183)**

KLEIDOGRAPH

The New York Institute for the Blind was responsible for this 1894 machine for embossing cells on paper inserted into the back of the instrument and winding itself around a drum. It had twelve keys and a space bar.

Rare, but blind machines tend to have limited commercial appeal, hence a price tag barely into three figures is about right.

KNEIST

An index machine of blatantly Hall Type-Writer inspiration was manufactured by Wunder & Kneist of Hannover in 1893 after a design (one can hardly call it an invention) by Otto Ferd. Mayer and J. Funcke of Berlin. The index was curved top and bottom and the carrier curved on both sides, instead of square as on the Hall. The rubber type was pressed against the paper by means of a separate arm, but otherwise there was little to separate the Kneist from the Hall. A model for the use of the blind was offered in 1901.

Well into four figures.

KOSMOPOLIT

A handsome product of the German sewing machine manufacturer Guhl & Harbeck who were already in the typewriter market with the Hammonia, both machines fitted to bases of early sewing machine inspiration. Unlike its predecessor, however, the 1888 Kosmopolit, invented by J. C. Koch, was of swinging sector design with a curved letter scale and a selector which was pressed down for printing, correct alignment assured by teeth on a comb of substantial proportions. The entire sector and selector housing was raised for paper insertion and moved a space at a time for printing. Upper and lower case, inking by pads. Also marketed as the Cosmopolit and Cosmopolitan.

Prices as high as $15,000 have been paid for examples of this highly desirable machine, but around $10,000-12,000 is more realistic. **(184)**

KRATZ-BOUSSAC

A machine by this improbable name is described as having a circular index with the raised letters around the periphery of a disk, pad inking, and a handle for selection and printing.[81] It sounds like the sort of machine Robert Ingersoll might have mass produced (cf. Dollar), or possibly the similar Herrington, and it was in all probability one of these, marketed under the name of a local retailer.

A couple of hundred, at the outside.

LA FRANÇAISE

A toy typewriter consisting of a circular index mounted obliquely over a thin roller and travelling along a rack as the index is pressed for printing. Contemporary with the Simplex and other similar machines, it differed just enough to pass patent obstacles. None of the machines appears to have been marked with either name of manufacturer or other identification, no doubt in order to facilitate its marketing under different retailers' labels, although one example, with a porcelain knob, is marked simply 'Made in Germany.' Identification of the machine was made possible only after the author found an example in its original presentation case, complete with small sized envelopes and writing paper, pen and inkwells, as well as instructions for the use of the machine, and the name of a distributor: N. K. Atles, Paris.

Might possibly fetch $100-200 or so, on a good day.

The machine was often dismissed in contemptuous terms in early literature—Martin brushes it aside as nothing more than '...ein nettes Spielzeug für Kinder und Erwachsene' (a cute toy for children and adults) but this is quite categorically incorrect.[63] The Lambert was made and sold extensively as a serious machine and documented evidence abounds that it was used for serious purposes for decades after it was first introduced.

The author feels it his duty, without further comment, to draw the attention of Mediterranean (and other) cooks and collectors to the manufacturer's advice on page 31 of the 1901 Lambert instruction booklet, which warns owners *not* to use olive oil for lubrication. 'Sperm is as good as any,' says the booklet, 'and getable everywhere.'

Excellent value at $700-1,000, or so. **(Front cover)**

L. C. SMITH

A conventional upright manufactured from 1904 by L. C. Smith Bros Typewriter Co. of Syracuse, New York. The Smith brothers (Lyman C., Wilbert L., Monroe C., and Hurlbut W.) had previously been responsible for the Smith Premier before it was engulfed by the Union Typewriter Company. The principal feature of the machine was a fixed carriage and moving type basket for change of case, but in competing against its better known 'blind' rivals it placed greater emphasis on its visible typing.

'You are Right Side Up; Why write Upside Down?' asked the blurb.

The machine was a commercial success from the start and production continued through numerous minor model changes. In the mid 1920s the company merged with the Corona Typewriter Company and their products were eventually marketed under the famous Smith-Corona label.

Collectors appear to like it, so prices tend to reflect demand in the $300-or-so range. **(185)**

LEO JOSEPH

A notoriously inaccurate source lists a single key machine by this name.[12]

LEVESQUE

Idem.

LIGNOSE

Yet another munitions manufacturer which diversified into typewriters was the Berlin firm of A. G. Lignose which produced a three row keyboard front stroke machine for a few months in 1924 before surrendering to better and cheaper four row competitors. Locating the ribbon spools on the sides of the machine instead of on top was one of its significant if far from original features.

$300-400 seems about right, although higher prices are known to have been paid.

LILIPUT

Justin Wilhelm Bamberger and Deutsche Kleinmaschinen-werke of Munich manufactured this simple circular index machine from 1907. Letter selection was performed by means of a knob on top of the device which was pressed down for printing on a anvil beneath. Roller inking. The paper was first curled inside the machine and exited between two feed rollers as typing progressed—a clumsy arrangement but by no means unique to this instrument. Several models were produced, each offering additional features and culminating in the Express which sported such attractions as upper and lower case and a space key. It was also marketed under the name Gnom.

LAMBERT

This darling of typewriter and phonograph enthusiasts is a quite remarkable machine, all the more amazing, since despite its peculiarities and its relatively late date, it sold in the many thousands the world over. Frank Lambert of New York developed the exotic design over a period of seventeen years, with the earliest patent dated 1884 and its eventual commercial materialization twelve years later. 'The least machinery yet for a typewriter,' boasted the manufacturer. '...small and compact—7 x 7 x 11 inches, about 5 lb, 140 parts—no nest of long levers. It is as if the types were on the tips of your fingers...'

Typing was performed by pressing down on the desired letter fitted on a rocking circular 'keyboard,' thereby tilting it so that the index of lenticular section attached to the bottom of the housing swung the desired letter above the printing point. Further pressure brought the character into contact with the paper, and releasing the pressure returned the index to its original position. Correct alignment was achieved by means of slots on the inside of the index and inking was by pad, of similarly curved profile, with which the characters were permanently in contact. Spacing came by pressing on the centre of the index, and upper and lower case by means of a lever on the left. The paper had first to be clamped under a bar running the length of the platen around which it then proceeded to wind itself as the page was being typed, requiring flattening out after completion.

It was in every way a remarkable invention. First manufactured by the Lambert Typewriter Co. of New York and on the other side of the Atlantic by the Gramophone Co., which changed its name to Gramophone and Typewriter Co. in its honour and marketed it from 1900 to 1904. Sidney Hebert persevered with it in France for a little longer and a German subsidiary also marketed it in that country. British and Continental examples had the name in raised letters in the casting on the base while on US machines the base is smooth and the name is applied in plain letters. A single mutant has surfaced in the United States with the name 'Butler' in place of Lambert, and with a serial number approaching 7,000. European examples are not recorded with serial numbers quite as high as that.

204. Oliver 3 (Courtesy of This Olde Office, Cathedral City, California)

Improbably enough, you are unlikely to get much change out of five grand. **(186)**

LINOWRITER

A double keyboard Smith Premier No 10 re-modelled by Empire Type Foundry of Buffalo, New York, so that the keyboard matched that of Linotype machines.

A rare mutant, so expect prices to reach four figures.

LOGOTYPE

A shorthand machine printing in upper case letters on paper tape, invented by Edna Robenson of Atlanta in 1923 and produced by the Atlanta Model Machine Co. the following year.

Rare, but shorthand machines tend to be less collectible than straight typewriters and therefore command lower prices. A price tag should not read more than a couple of hundred dollars or so.

LONG

Despite its name, this was a tiny shorthand machine with four keys, small enough to be carried and used in one hand. Subject of a 1900 U.S. patent in the name of Eugene McLean Long, it was manufactured the following year. It printed a shorthand code of strokes on paper tape by the use of up to four keys, with three positions per key. Sometimes referred to under the names Long and Callaghan, also McLean Long and Callaghan.

Cute, and rare, so expect to pay up to four or five hundred.

LONGINI

H. E. Longini of Brussels manufactured this type wheel device in 1906 for printing signs in large letters. Indicator for letter selection, roller inking, printing by means of a key on top of the machine.

Rare. If you find one for less than four figures, prepare for larceny charges!

LORD BALTIMORE

Toy typewriter manufactured by Baumgarten and Co., Baltimore. Circular index, operated by a knob, positioned horizon-

205. Olympia (Courtesy of Bonhams, London)

tally above a platen and brought down into contact with the paper by means of a separate printing lever.

Likely to set you back a few hundred.

LUDOLF

1930s toy typewriter of German origin.

A recorded $700 or so has been paid for one.

MANHATTAN

One of several attempts at re-animating the famous Remington Model 2 after it had been superseded and its patents expired. The Manhattan Typewriter Co. of New York was responsible for this 1898 machine which was produced in two very similar models designated 'A' and 'B,' and which enjoyed reasonable sales for some years before production was halted. An attempted re-animation in 1905 by the Blake Typewriter Co. of Newark, New Jersey, was unsuccessful and ended in the hands of the receiver four years later.

$600-800 for this relatively rare mutant.

MANOGRAPH

Toy typewriter of index design made by Gebr. Heilbuth, Hamburg, in 1906. Toy typewriters have a dedicated specialist following, so prices in the hundreds...or even thousands...may seem disproportionately high for machines like this.

MAP

French front stroke upright produced from 1921 in several models by Manufacture d'Armes de Paris to utilize industrial capacity made available by the cessation of hostilities.

Three figures would be altogether too high a price to pay for this relatively dull machine, but it might be nice to couple one with a Typo, the French version of the Imperial (1), also made by MAP.

MASKELYNE

John Nevil Maskelyne and his son invented this fine and now rare machine in 1889, patenting it the following year. It appeared in several models, none of which proved commercially successful despite the fact that the designs could boast some sophisticated features.

First two models were of grasshopper design with the horizontal type bars fanning out from the printing point. Inking by pad. Three row keyboard with double shift. Full differential spacing was achieved by the use of four universal bars to advance the platen a single space for diphthongs or two, three, or four spaces according to the width of the letter selected.

206. Pearl (Courtesy of Tom Fitzgerald)

1897 saw the introduction of the third model, also known as the Maskelyne Victoria, which featured one of the most remarkable actions of any typewriter ever produced. The horizontal type bars, with the type facing upwards and resting against an inking pad which faced downwards, performed somersaults on their way to the platen, after first lowering themselves from the pad. Once observed, never forgotten.

Price? First of all think five figures, then work out how high.

MASSPRO

Three row keyboard, double shift machine of front stroke design invented by George F. Rose who, with his father, had previously developed the Standard Folding. Manufactured from 1932 by the Mass Production Corp. of New York, from which the machine took its name.

Relatively rarely seen, but of no intrinsic interest other than the inventor...or rather his father. If it is priced higher than $20-30, forget the father.

McCOOL

William A. McCool invented this type wheel machine with three row keyboard, protected by a patent filed in 1903, granted seven years later, and assigned to the Acme-Keystone Manufacturing Co. of Beaver Falls, Pennsylvania, who produced it in small numbers.

Five figures, for sure.

McLOUGHLIN BROS

A $10 typewriter promoted as the 'cheapest in the market,' with applications ranging from school children to business men and a claimed performance comparable to machines costing up to ten times its price, this circular index device using rubber type attached to a disk which travelled along a rack above a stationary platen was marketed by New York toy dealers McLoughlin Bros. from 1884.

It might now be rare, and desirable, but the prices of up to $10,000 paid in the past for this machine are beyond this writer's understanding, so no further comment.

MENTOR

Four row portable of oblique front stroke design, manufactured in Germany by Metallindustrie A.G. from 1909. Also marketed under the names Thuringia and Monofix.

Prices have been registered in the region of $1,000, and then some.

207. Peerless (Courtesy of DeWitt Historical Society of Tomkins County, Ithaca, New York)

208. People's (Courtesy of This Olde Office, Cathedral City, California)

209. Perfected Type Writer No. 4

MERCEDES

Conventional upright manufactured from 1907 in Germany by Mercedes Büromaschinen-Werke A.G. Many models, including portables and electrics. Also sold under a variety of names in different countries, including Protos, Cosmopolita, and Drake.

Many models, but if they reach three figures they are already too expensive.

MERCURY

This small three row keyboard machine with a type wheel offset to the extreme right to make typing fully visible to the operator was protected by a British patent granted to F Myers in 1887 and manufactured for a short time by the Mercury Typewriting Co. of Liverpool. Printing was performed by means of a separate key on the right which raised the platen up against the type wheel. Roller inking.

Very, very rare, so think five figures.

MERRITT

A remarkable machine of up stroke linear index design patented by M. G. Merritt of Massachusetts, in 1890. Produced by Merritt Manufacturing Co. of Springfield, Massachusetts, and later by Lyon Manufacturing Co. of New York, the machine featured a row of individual type slugs in a carrier sliding under the platen. A small knob served to select the correct letter on an index and pressing it down into the corresponding slot brought type plunger in contact with paper. Roller inking. The platen was hinged at the rear and could be raised to make typing visible to the operator. It seemed an unlikely design, yet many thousands of these machines were made.

Not a rare machine, but prices up to $1,000 and more are regularly paid, reflecting its appeal rather than its rarity. **(188, 270)**

MERZ

German front stroke portable with four row keyboard produced from 1926 by Merzwerke of Frankfurt. Sold in France under the name of Concord.

$20…$30…even $40, perhaps.

METEOR

Popular front stroke portable with three row keyboard and double shift marketed under a diversity of names over a fourteen year period starting in 1911. Produced by Sachsische Strickmaschinenfabrik Meteor until 1922, then by Vasanta Schreib- und Strickmaschinenfabrik A. G., both of Dresden. Production came to an end in 1925. Variously marketed in different parts of the world under the names Berolina, Doropa, Forte-Type, Janus, Pagina, Vasanta, and Wilson.

Up to $100 or so.

MICHELA

No example of a fine shorthand machine designed and built by the Italian brothers Antonio and Giovanni Michela appears to have survived, despite the success the instrument enjoyed. Developmental work began well before 1862, when the first model was eventually completed, with public presentation the following year, but it was not until 1876 that the first patent was granted, a second one following two years later.

From a distance, the machine might easily have been mistaken for a harmonium with its turned wooden legs and piano keyboard of ten keys (six white and four black) for each hand, separated by a roll of paper tape 44 mm wide of which 300 metres was sufficient for seven hours of dictation. Keys were played in chords and printed symbols of the component sounds according to the inventor's own system. Thumbs of each hand operated two white keys, while the other fingers were used for one white and one black key each, permitting the hands to be kept virtually still while only the fingers moved. Inside the case, the action of the keys was transmitted by levers towards the tape in the centre. The symbols were originally embossed on the paper tape but a ribbon was later used to make them more visible.

Contemporary reports speak highly of the machine which permitted stenographers to record dictation at the speed of speech for three hours '*sans fatiguer*,' instead of the usual quarter of an hour, and permitted them to read back what they had written without the customary difficulty and hesitation. The Italian Senate and Chamber of Deputies used it 'exclusively' from 1880; Michela's French agent is recorded as having demonstrated it to the French Senate, Chamber of Deputies, and Municipal Councils from 1881. At the same time, courses for stenographers using the instrument and its system were widely advertised and apparently well attended. Despite all this, no example of the machine appears to have survived, which is the more surprising since its use continued well into the 20th century when descriptions and reports of it were still being published in the present tense. **(13, 14)**

MIGNON

Just a simple glance at this remarkable machine will suffice to explain why it is so popular today with connoisseur and dilettante alike but no amount of insight can explain why such a quaint but anachronistic device should have been as popular as it was in its day, given its relatively late arrival on the scene.

But popular it was, and for many decades at that, made and sold in its hundreds of thousands most of which (or so it seems) have managed to survive to this day.

Dr. Friedrich von Hefner-Alteneck was responsible for designing the machine in 1903 for the German manufacturing concern Allgemeinen Elektrizitäts-Gesellschaft of Berlin (AEG). An indicator operated by the left hand and connected to a shaft geared to a type sleeve was suspended above a rectangular index; the indicator was simply pointed at the desired letter and the corresponding character on the type sleeve was brought down onto the platen by means of a separate key operated by the right hand. Ribbon, with the spools on either side of the type sleeve housing. A separate space key was provided, with a further key for back-spacing on Model 4. Other than that, and other minor alterations such as the shape of the base and a change from hard moulded cardboard to metal for

210. Perfected Type Writer No. 4, typing sample, 1912

Telephone No. 5555 Avenue (5 Lines)
Telegraphic Address, Wyckoff, London.

ABERDEEN. BELFAST. BIRMINGHAM. BRADFORD. BRIGHTON. BRISTOL. CARDIFF. COVENTRY. DUBLIN. DUNDEE.
EASTBOURNE. EDINBURGH. GLASGOW. HANLEY. HUDDERSFIELD. HULL. INVERNESS. LEEDS. LEICESTER. LIVERPOOL.
MANCHESTER. NEWCASTLE. NORWICH. NOTTINGHAM. PRESTON. SHEFFIELD. SOUTHAMPTON. YORK.

BY APPOINTMENT TO HIS MAJESTY KING GEORGE V.

CONTRACTORS TO
HIS MAJESTY'S GOVERNMENT.
THE INDIAN GOVERNMENT.
AES/P

Remington Typewriter Company, LTD.

100, Gracechurch Street,
London, E.C.

October 8th, 1912.

A. W. Moore Esq.,

   Calverton Lodge, Stony Stratford, B u c k s.

Dear Sir :

   We are in receipt of your favour dated the 4th instant
which is to hand to-day, and note that you require some
adjustment made to your typewriter.   With this object in
view we have sent your enquiry to our local office (15, King
Street, Leicester) requesting them to give the matter their
prompt attention; they will be able to give you all the
information you require.

   Assuring you of our best attention at all times, we are,

       Yours faithfully,

           REMINGTON TYPEWRITER CO. LTD.

---

211. Perfected Type Writer No. 4: '...you require some adjustments...'
letter from Remington Typewriter Company, 1912.

the type sleeve housing, the essential design was remarkably durable and enjoyed few changes over the decades. Model 2 dates from 1905, Model 3 from 1913, and Model 4 from 1923. The design of the machine lent itself particularly well to all applications for which interchangeable type was required and literally dozens and dozens of different type wheels with their corresponding indices were offered, including ones for the blind. An abortive attempt was even made to electrify the carriage at a time when it became fashionable to fit electric motors to the outside of existing manuals for this purpose.

Variously marketed under the names AEG, Yu Ess, Special, Stallman, Heady, Stella, Eclipse, and Plurotyp. A thinly disguised version was manufactured in Czechoslovakia in 1936 and named Tip-Tip.

relatives, as was the 1927 Helma. Several models of the Minerva were marketed, all similar. Also served as the basis of a model called Polyglott.

$100-200 should be about right.

MINERVA (2)

German machine for the blind with six keys and a space bar, made in Leipzig in 1928.

About the same as previous.

MITEX

Mitex Schreibmaschinen GmbH produced this three row front stroke portable with double shift in 1922, changing its name to Tell and the company to Tell Schreibmaschinen GmbH the following year. Manufacture was discontinued after a few years but the design was resurrected in England in the 1930s as the

212. Perkins Braille Writer

Mignons were generally black, although other colours are known and an example finished in red recently appeared and caused quite a stir.

Despite its apparently anachronistic design, the Mignon continued to be manufactured and marketed until well into the 1930s at the same time as the company was producing a conventional upright, named after its initials: AEG.

Despite appearances, Mignons are ubiquitous and therefore look far more expensive than they are: uninformed vendors offering one for sale will usually swear it is the first machine ever made. Only Germans pay funny money for rare examples in colors other than black—the rest of the world is relatively unimpressed. Expect to pay from, say, $150 to $500, depending on model, condition and completeness. Colours other than black fetch four figures. **(47, 189, 190)**

MIKRO

A tiny machine by this appropriate name and measuring a mere 6 x 7 x 3 inches overall (without the carriage) was manufactured in 1925 by Paul Schütze GmbH of Dresden. Two row keyboard of sixteen keys with double shift, printing only upper-case.

Rare, so price is likely to be correspondingly high. Slip comfortably into four figure mode, and hope for the best.

MINERVA (1)

Conventional front stroke upright with four row keyboard first introduced in 1909 by the Deutschen Schreibmaschinenwerken Hövelmann, Kührt & Bollendorf, later after the First World War by Minerva Schreibmaschinenfabrik GmbH of Nürnberg. It was a highly successful design, marketed in different countries under a multitude of names, but its success was relatively short-lived and it succumbed in the mid-20s. The Italian Fidat and the later Fontana were close

Bar-Let, produced by the Bar-Lock Typewriter Co. of Nottingham which was also responsible for the machine which gave the company its name.

Relatively rare, which explains why $700 or so was paid for one some years ago.

MOLLE

Collectors anxious to find a Model One and a Model Two of this lightweight front stroke machine with three row keyboard will be disappointed (or relieved, perhaps) to learn that the Molle No. 3 is the first of the species. The reason is that the inventor, John E. Molle of Wisconsin, was the author of two previous designs which were aborted prior to birth. The No. 3, patented in 1913 and manufactured by the Molle Typewriter Co. five years later, was designed to facilitate servicing and repair, but its chances of success were slim to begin with, and were not helped by the death of the inventor, the company going into liquidation in 1922. Attempts at re-animation under the name Liberty, by the Liberty Typewriter Co. of Chicago, were short-lived.

A pleasant machine in the $200-400 bracket. **(191)**

MONARCH

Conventional front stroke upright machine manufactured from 1904 by the Monarch Typewriter Company of Syracuse, New York. From 1907 or so, it became one of the members of Union Typewriter Co., the trust composed of Remington, Smith Premier, Caligraph, Yost, etc. Several similar models were produced before it was eventually phased out, together with the other makes, in favour of the Remington. A three row portable model called Monarch Pioneer was introduced in 1920. **(192)**

$100-200 is good value for money for the upright Monarch Visible **(193)**. The portable is worth considerably less.

213. Pettypet (Courtesy of Christie's South Kensington, London)

MORRIS

Robert Morris of Kansas City was the inventor of this small machine which used rubber type beneath a housing pivoted at one end, with a corresponding index and indicator above. A knob on the housing operated by the right hand selected the desired letter on a plate (printed on some, enamelled on others) with five curved rows of letters corresponding to those on the type below. Inking by pad. Pressing the knob produced the impression on paper on a platen and advanced the index housing a space to the right. Manufacture began in 1886 when the patent was only 'applied for'—it was not granted until January of the following year. Manufacturer was The Hoggson & Pettis Manufacturing Co., New Haven, Connecticut. The upper case only machine was offered in two models, No. 1 having smaller type than No. 2, but despite this few were made: an example previously in the author's collection was serial number 111. The original instruction booklet refers to it as both a 'type writer' and a 'type-writer.'

Rare machine. A price approaching $20,000 was paid for one some years ago but $10,000-15,000 is probably more sane. **(194)**

MOYA

A prolific and versatile Spanish-American inventor (of *inter alia,* the 'Moya violin') living in Leicester, England, by the name of Hidalgo Moya ('commonly known as Dalgo Moya,' according to a provisional 1901 patent) was responsible for the type sleeve machine which bore his name. The first model was fully covered by a patent dated 1902 and was marketed the following year, with three row keyboard and ribbon, the spools located rather clumsily between keyboard and type sleeve. This model was not a commercial success, despite its more than competitive' price of five guineas, and was replaced in 1905 by Model 2, still with three row keyboard and type sleeve, but with the movement now concealed beneath a cover and the ribbon spools located on the sides. This model enjoyed somewhat more success both at home and abroad; in France it was sold under the name Baka 1 from 1908 by Manufacture Française d'Armes et Cycles de Saint-Etienne, who subsequently marketed the Imperial under the name Typo, and in Germany it was manufactured under licence by Seidel and Naumann of Dresden and marketed under their name 'Ideal.' A third model is known to have been developed in 1907 but it was not marketed because by then Moya had secured financial backing and formed the Imperial Typewriter Company the following year, launching a completely new design called Imperial and dropping the Moya altogether.

According to contemporary sources, the resourceful 'Dalgo' Moya was 'the greatest authority on the mechanism of the typewriter in Europe,' and later went on to become Stearns Visible agent for United Kingdom.

Prices range from over $3,000 or so for the first model down to $1,500 for the second. **(93, 195, 196)**

MOYER

Conventional front stroke upright designed by Emmet G. Latta of New York in 1913 and launched briefly under the above name before becoming the more familiar Bar-Blick, manufactured by the Blickensderfer Manufacturing Co. The machine became the Harry A. Smith when the Blickensderfer factory closed in 1917.

A hundred or two should seal a deal for it.

MUNSON

A horizontal type sleeve mounted parallel to the platen and struck by a hammer from behind the paper was the principal curiosity of this interesting machine patented by Samuel John

214. Picht (Courtesy of Sotheby's, London)

215. Picht **(c)**, Stainsby Wayne **(l & r)** (Courtesy of Christie's South Kensington, London)

216. Pittsburg Visible 10 (Courtesy of This Olde Office, Cathedral City, California)

Seifried in 1889, with two-thirds assigned to Fred and Louis Munson of Chicago whose Munson Typewriter Co. introduced it to the market the following year. A three row keyboard was used, and interchangeable type sleeves and matching sets of interchangeable keys were offered.

The machine was well-received and met with considerable success. A slightly improved model was offered in 1897 but a year later the enterprise was bought up by Edgar A. Hill and the name of the company changed to the Chicago Writing Machine Co., and that of the machine to Chicago **(135)**.

$1,200-1,500 for the first model Munson, slightly more for the improved version, although much higher prices are known to have been paid **(197)**.

MURRAY

Printing telegraph, using perforated paper tape and a modified Bar-Lock typewriter, introduced in the United States in 1899 by an Englishman called Donald Murray. Early models were described rather unflatteringly as coffee mills or sausage machines due to the fact that a handle had to be turned to operate the instrument, but improvements in operation and design soon followed and the machine eventually enjoyed considerable commercial success in many countries, from Russia to India.

Rare. In fact, very, very rare so that can only mean four figures.

NATIONAL (1)

A man called Henry Harmon Unz was granted a patent in 1889 for this up stroke machine with a curved three row keyboard manufactured by the National Typewriter Co. of Philadelphia but the design belongs to Franz X. Wagner of Underwood fame—his 1885 patent for a similar machine was assigned to Unz.

$4,000? $4,500? Maybe even a little more for a pristine example. **(198)**

NATIONAL (2)

A conventional front stroke portable with a three row keyboard and double shift. The Model Two was the first on the market in 1916/17, Model Three the following year, and the Five in 1920. Manufacturer was Rex Typewriter Co. It was sold under the following names: Crown, Express, National Portable, and Portex.

$150-200 should be plenty. **(150 [r], 218)**

NATIONAL (3)

Shorthand machine produced in 1916 by Ward S. Ireland who was previously responsible for the Stenotype to which the National bore more than a casual resemblance.

One or two hundred should be enough.

NEW AMERICAN

Identical to the International (1), this product of the New American Manufacturing Co. may possibly have been the export version: an example previously in the author's collection had the name-plate in French, listing French patents. Date of manufacture is 1890 or possibly slightly earlier, based again upon Lee Burridge's 1885 patent.

$1,500-2,000.

NEYA

Arpad Krejniker invented this three row front stroke machine with double shift, and A. Ney of Berlin, who had begun making the Helios Klimax at the beginning of the First World War, manufactured it briefly in 1925.

$200 or so should seal it.

NIAGARA

The Blickensderfer Manufacturing Co. was responsible for this machine which was largely assembled from Blickensderfer parts but which dispensed with a keyboard in favour of a vertical circular index with the type wheel at the other side of the indicator shaft. It was an inexpensive machine, designed to compete at the lower end of the market. Precise date of manufacture is uncertain but was around 1902 or so. Also marketed under the name Best.

Rare. Think $2,000-3,000, to be safe.

NICKERSON

Rev. Charles S. Nickerson's first effort was an 1897 patent for an up stroke machine which never left the drawing board, but he is better remembered for his later efforts to manufacture a remarkable vertical platen typewriter based upon ideas first evolved in 1899 by Walter Hanson of Milwaukee, whose front stroke machine with its four row keyboard which typed around, not along, the vertically-mounted platen apparently passed from the hands of the inventor, upon his untimely death, to those of a Rev. Lee and eventually to Nickerson. The patent date is 1909.

The Nickerson Typewriter Company was formed in 1907 to bring the design to manufacture, but despite the expenditure of large sums, few machines were made and one, in the Milwaukee Museum, appears to be the only survivor.

217. Pittsburg Visible 12 (Courtesy of This Olde Office, Cathedral City, California)

If there is another survivor anywhere, it is likely to make five figures.

NIPPON

The enormous complexity of designing a typewriter to handle thousands of characters—according to Yamada Hisao, 1,500 to 3,500 of these are required for composing even ordinary documents, although the largest Japanese dictionary records about 50,000 characters—was tackled in several ways by early inventors.[94] Some machines **(251, 252)** had multiple trays containing individual type which was picked out one by one for printing and then replaced before the next characters were selected. Later models such as the Ôtani **(253)** had row after row of ideographs on rubber belts and was briefly manufactured in 1940. A nifty portable for use by the army **(254 a & b)** was produced from 1931 or so.

Manufactured from 1915 by Nippon Writing Machine Ltd (Nippon Syoziki Syokai) of Tokyo, later Nippon Typewriter Co. Any surviving examples are likely to fetch sophisticated prices.

218. Portex 5 (Courtesy of This Olde Office, Cathedral City, California)

NOISELESS

An important attempt at producing a silent typewriter was made by the inventive Wellington P. Kidder who is perhaps best remembered for the highly successful thrust action design to which his Christian name was applied. Together with a Canadian called C. C. Colby, who was largely responsible for the world-wide success of the Wellington-Empire-Adler format **(157)**, Kidder began designing a silent machine as early as 1896, intending to call it, appropriately, the 'Silent' although it was not until 1912 that the first Noiseless machines were produced. They were not, of course, noiseless at all but merely less noisy than others; the secret of the design was to bring the type face almost to the paper and push it the final little piece by means of an overthrow weight, rather than have it strike by its accumulated momentum.

Production remained limited until the early bugs had been ironed out of the design which took until 1917 or so with the launching of the Model 4. A portable was added to the range four years later and the Remington Company took over in 1924, a year or so after the introduction of the fifth model.

Model 6, sporting a conventional four row keyboard, and now known as Remington Noiseless, followed in 1925 and all subsequent models used both names. Prices likely to range from $400-600, depending on model. **(84, 200 [I])**

NORDEN

Danish conventional upright, first introduced in 1918 by A. B. Nordiske Skrivemaskinefabriker of Copenhagen under the name Skandia, later Baltica, then, with concurrent changes of manufacturer, as Norden in Denmark and Halda-Norden (eventually simply Halda) in Sweden. In 1938 the name was changed yet again to the now more familiar Facit.

No doubt early models are now sought after in Scandinavia, but over here if someone asks you for a three figure sum for one, ask your attorney to press charges.

NORICA

German oblique front stroke machine introduced in 1907 by Kührt and Riegelmann GmbH, Nürnberg and produced in limited numbers through minor design changes but retaining the oblique front stroke concept. Production was taken over two

219. Postal (Courtesy of This Olde Office, Cathedral City, California)

years later, by Deutsche Triumphfahrradwerke A. G., the German subsidiary of the British Triumph motorcycle concern, who marketed it as a conventional front stroke machine under its own name.

A few hundred dollars should do it, with spare change left over.

NORTH'S

This desirable down stroke machine, so often incorrectly referred to without its possessive suffix, was named after Lord North who financed its brief manufacture until his untimely death. Morgan Donne and George Cooper were responsible for this 1890 design, the former of the two having played a part in the short-lived English with which the North's shared some features. Four row keyboard, with type bars positioned vertically behind the platen thereby necessitating the use of open framework for coiling the sheet of paper since it could not enter and exit as it would have done on a front or up stroke machine: this design feature may well have hastened the machine's demise. Despite the positioning of the type bars behind the platen, typing was not immediately visible, being

220. Remington Standard 7

covered initially by the ribbon, but this could be lifted to reveal what had just been printed.

The North's Typewriter Manufacturing Company of London was responsible for placing the machine on the market in 1892. It also appeared in France under the name Nord.

$4,000-5,000 may no longer be enough for a good example. **(79, 201)**

NOTOTYP-RUNDSTATLER

One of the special-purpose adaptations of the German Archo typewriter which its manufacturer Carl Winterling developed in 1936, this one in conjunction with a man called Rundstatler, for the purpose of typing musical scores. A second model called Melotyp was later introduced.

Few are known, but a few hundred dollars should be ample.

NOUVEAU SIÈCLE

French circular index machine similar to the Simplex introduced by G. Meyer in 1912. Also sold under the name Meyer. If the price reaches three figures it is high.

NOWAK

A dual purpose machine with interchangeable type wheels suitable for typing or embossing, this machine for the blind was designed by an Austrian called Nowak and manufactured by Sczepanski and Co. of Vienna in 1902, with a US patent granted the following year.

ODELL

L. J. Odell of Chicago designed this distinctive linear index machine in which the index travelled across the platen at right angles, with the index housing pivoted at the far end. A small projection on the index held between thumb and fore-finger selected the desired letter and was then pressed down for printing. Roller inking. On models offering upper and lower case, a separate key was used to rock the index to the desired position. The patent protecting the design was filed in 1887 and granted two years later, although a contemporary brochure

claimed that the company had been in existence since 1877. In all, four models were produced.

The essential features of this machine were identical to those used on the Sun, manufactured by the Sun Typewriter Co. of New York after a patent granted in 1885. On the strength of this dating, the inventors of the Sun might well claim priority, which might also help explain why Odell promoted the 1877 date although the very fact itself that patents were granted to both inventors for virtually identical designs remains mystery enough.

Other machines labelled America, New American, and International were also manufactured to these specifications and survived in the market place until well into the first decade of this century.

Spectacular design, which might induce anyone to pay over the odds. The machine is rare, but not all that rare…$1,000-$1,500 is plenty. **(202)**

ODOMA

Conventional German upright first named simply Odo, introduced in 1921 by Odo Maschinenfabrik GmbH of Darmstadt, Germany. Production continued until 1937. Also sold under the name Blickensderfer-Odoma.

Somewhere between $1 and $150 ought to do it.

OFFICIAL

A small hammer at the back of this machine was used to strike paper against type wheel, a feature already widely popularized many years before 1898 when C. E. Peterson of Minneapolis designed the machine which he placed on the market three years later as the Official. Three row keyboard with double shift, ribbon inking. It was a simple machine of small dimensions and few parts, but it enjoyed only a limited life.

Very rare, so if one appears, throw your pocket-book at the owner, close your eyes and tell him to help himself.

OFFICINE GALILEO

An Italian index machine about which little is known and only one example appears to have survived: this is now in the Museum of Science and Technology in Milano. Dated 1904, the design uses a rectangular index, with indicator, and a cylindrical platen of conventional proportions.

OLIVER

Inverted U-shaped type bars banked to the left and right of the printing point are the characteristic features of this heavy cast-iron machine built to last forever…and many of them doubtless will, if their tendency to rust can be restrained. The Rev Thomas Oliver took four years to develop his first machine for which he was granted a patent in 1891, but the de-

222. Remington Noiseless 6 (Courtesy of This Olde Office, Cathedral City, California)

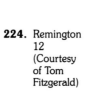

224. Remington 12 (Courtesy of Tom Fitzgerald)

sign featured a segmental comb with roller inking and a hammer striking the type to produce the impression. Elements of this design remained with him when he completed a totally different concept using the lateral type bars which give the Oliver its characteristic profile: this design was patented in 1894 and the first machines appeared the same year, but full-scale production by the Oliver Typewriter Co. of Chicago only began two years later with the Model 2. Model 3 followed in 1898, by which time the reputation of the machine was assured. Major models were the 5 in 1906 followed by the 7 in 1914, the 9 in 1916, and the 11 in 1922.

A feature of the early machines was a projection on either side of the housing which was intended to make carrying easier. These projections were flat on early machines and of marginal benefit; thereafter they were raised and curved and did in fact greatly facilitate carrying the heavy ungainly Olivers, before being phased out altogether by Model 11.

The design proved so good that model changes were limited to such details as back-spacing, ribbon design, tabulator, and so on, and production of this make reached telephone-number proportions, the more so since their sturdy construction made them an obvious choice for the armed forces before and during the War. Large numbers, in war-time finish, survive to this day.

Production in the United States ceased in 1928 but continued in Europe for considerably longer. Model 15, labelled British

Oliver, continued until 1931, when the design finally succumbed to the demands for standardization, to be replaced by a conventional front stroke portable with four row keyboard which was manufactured under licence in a number of European countries.

The make was also marketed in Austria under the name Courier and in Germany as the Fiver and the Stolzenberg. An example labelled Jacobi has also been discovered.

Only the early models are worth up to four figures. Later models are common and ought to be buyable for a hundred or two, some even less. Hold out for examples in good condition because most Olivers (like some of us humans!) tend to show their age more than most. **(203, 204)**

OLIVETTI

Camillo Olivetti of Ivrea, Italy, launched the first model of this important machine in 1911 and the manufacturer is of course still very much in evidence to this day. First model was of conventional upright design, with a stepped four row keyboard reminiscent of early Royals and Smith Premiers. Production of angular uprights continued until the 1950s and before that date, the features of successive model changes showed only slight modification. Model M 20 was similar to the first model and retained the stepped keyboard, which was abandoned in the subsequent Model 40 of 1930 in favour of a strictly conventional design. Post-war models showed progressive softening of lines.

223. Remington Noiseless 8 (Courtesy of Bernard Williams)

225. Rem-Sho (Courtesy of Bernard Williams)

226. Rex Visible 4 (Courtesy of This Olde Office, Cathedral City, California)

The first model may be worth a hundred or so, on a pleasant day. Later models progressively less.

OLYMPIA
The conventional upright by this name began its life as the AEG manufactured by Allgemeinen Elektrizitäts-Gesellschaft of Berlin which underwent several name changes before becoming the Olympia Büromaschinenwerke A. G. in 1936. However, the AEG Model 7 of 1930 already carried the Olympia logo and the appearance of the machine continued virtually unchanged until the 1950s.

Prices from a hundred, perhaps, right down to... **(205)**

ORGA
Conventional front stroke upright manufactured in numerous models from 1922, first by Bing-Werke A. G. of Nürnberg and, from 1932, by Orga A. G. of the same city.

Zzzzzz...if you pay three figures you've been suckered.

PACIOR
Wl. Paciorkiewicz was responsible for what he claimed was the first Polish typewriter which was a front stroke type bar machine with three row keyboard and double shift. Introduced in 1921, it was also marketed under the names Idea and Iskra.

Rare but who cares, apart perhaps from Polish collectors.

PALANTYPE
Shorthand machine of conventional design dating from the early years of this century.

Two figure sums ought to be plenty.

PARAGON
This previously unrecorded name has appeared on the top of a Remington Model 5.

You might pay a little more than Remington 5 price, if you collect pseudonyms—this name is rare.

PARISIENNE
A French circular index machine consisting of a wheel around the rim of which is the rubber type; pushing down on the selector arm causes the type to be pressed against the platen. Pad inking. Ernest A. Enjalbert of Paris was responsible for this 1885 invention.

Prices ought to be commensurate with comparable Simplex etc. examples.

PEARL (1)
'Errors, like straws, upon the surface flow;
He who would search for pearls must dive below.'

Had Dryden been writing about the typewriter by this name he could hardly have composed a more appropriate couplet. The confusion surrounding the Pearl, and its siblings the Peoples and the Champion, has still not been completely resolved despite the best efforts of students and collectors alike. At first glance, the confusion is easy to excuse because the machines are in fact very similar and the manufacturer himself may well have been partially to blame, for the three above-named machines, representing only minor changes one from the other, were introduced within a relatively short period of time and some examples may well have been modified from one specification to another during production itself.

They all print by means of a type wheel geared to an indicator selecting characters from a curved index. The variations, however, become apparent on closer scrutiny, both in the method of printing and, more particularly, in the inking systems.

C. J. A. Sjoberg of Brooklyn, New York, was the inventor, with patents dating from 1891 assigned to the Garvin Machine Co. of the same city. The Pearl was introduced around the middle of 1891 and its particular features are that the rubber type was mounted on only some 240° of the type wheel's circumference, printing only upper case, with inking by means of a roller on each side of the printing point. The type wheel was turned by a lever which also served for printing and, when 'space' was selected on the index, for spacing.

The Pearl enjoyed only limited sales, and production was soon abandoned. Very few examples survive.

Four figures—rarer than Peoples and Champion, so price is likely to be correspondingly higher. **(206)**

PEARL (2)
Even fewer examples survive of a substantial transverse linear index machine by this name, mechanically similar to the Sun and manufactured by Enoch Prouty of Chicago in 1887.

Price is going to be correspondingly higher than that of the Sun, so expect to find yourself more than half way to five figures...

PEERLESS
Up stroke machine with a double keyboard which was manufactured briefly from 1891 onwards by the Peerless Typewriter Co. of Ithaca, New York. It was virtually identical to the Smith Premier with which it fought an up-hill legal battle over patent infringement. Charles M. Clinton and James McNamara were responsible for the 1890 patent.

If you pay less than low four figures you have done well. **(207)**

227. Rico (Courtesy of This Olde Office, Cathedral City, California)

PEIRCE

This product of the Peirce Accounting Machine Company manufactured in 1912 consisted of a three row keyboard front stroke machine with a vertical flat paper frame.

Price is not likely to reflect rarity—this is the sort of 'find' you might be able to pick up for next to nothing.

PEOPLE'S (1)

A transverse linear index machine with characters on both sides of the index, (or, for upper and lower case, on the three sides of an index of triangular section), this machine was reported to have been invented by a man called E. Prouty and marketed in 1885.[46] A design of comparable conception was patented by H. J. Thomas the following year.

One example has surfaced, but so far that is all. Price must be commensurate with this sort of rarity.

PEOPLE'S (2)

Sister machine to the Pearl, with which it was possibly contemporary, the People's was the outcome of a patent granted to C. J. A. Sjoberg of Brooklyn in 1891, and manufactured from the same year by the Garvin Machine Co. of the same city. Type wheel design, with the name cast into the spokes of the type wheel. Letters on a curved index, upper and lower case, with a key for selection plus additional keys for printing and spacing. Printing was performed by rocking the platen forward against the type wheel.

Inking was by means of a roller which automatically replenished itself from a reservoir. This system was apparently unsuccessful, however, for it was soon replaced by the peculiar ribbon loop outlined in Sjoberg's 1891 patent: the loop was clipped to two hoops the upper of which rested on a carrier pivoted in the type wheel spindle while the lower, of slightly larger diameter, merely hung from the ribbon loop to prevent distortion. Movement of the ribbon loop was achieved solely by friction and vibration. This was clearly no better than the roller and reservoir, for both systems were soon superseded by the more sophisticated model marketed under the name Champion.

$1,500 or so should be sufficient. **(134 [r], 208)**

PERFECTED TYPE WRITER

One of the transient stages through which the Sholes and Glidden Type Writer briefly passed on its way to becoming the

229. Royal Portable (Courtesy of Sandy Sellers)

Remington. Originally called Type Writer, then Improved Type Writer (models up to no. 4 of this designation are known, and there were possibly more), followed by Perfected Type Writer (also numbered to 4) and finally Remington No 2, the simple designation Type Writer, which already by then had become generic, being finally dropped.

The machine illustrated, formerly in the author's collection, bears the serial no. 347. The correspondence came with the machine.

This model is very much a connoisseur's instrument and is a great deal rarer than a Sholes and Glidden, but not necessarily more desirable nor worth more on the open market. In any case, think in terms of no less than $5,000-10,000, to be on the safe side. **(209, 210, 211)**

PERFECTION

Type bar machine invented by the prolific Richard Uhlig who spawned some fifty or so designs of which thirteen, including the present 1906 machine, were actually manufactured. Details are sparse, and no example appears to have survived.

PERKEO

German version of the popular Standard Folding/Corona machine **(141, 249)**, produced from 1912 by Clemens Müller A. G. of Dresden after it had bought up the Viennese firm of Carl Engler which had been marketing the machine under the name Albus. In France it was sold as the Galiette.

Low three figures, no more.

PERKINS

Braille machine produced by the Perkins Institute of New York from 1900 onwards. First model had six keys plus spacer and used a conventional carriage but five years later saw the introduction of an improved model which sported what was to become the standard keyboard of machines for the blind.

Machines for the blind tend to be less hotly contested than straight typewriters...with a few notable exceptions...so the low three figures ought to be ample. **(212)**

PETTYPET

Small primitive type wheel machine with sliding indicators invented by an Austrian called Podleci and manufactured by the Archo factory around 1930. Ribbon inking. Also called Petty; Pettypet GmbH was responsible for the marketing. A slightly modified version using an indicator manipulated by

228. Royal (Courtesy of This Olde Office, Cathedral City, California)

230. Royal Portable (Courtesy of Sandy Sellers)

an independent stylus was later produced at the Stylotype Kleinschreibmaschinen GmbH and labelled Stylotype.

A couple of hundred ought to secure it. **(213)**

PHÖNIX

Type wheel machine invented by Wladislaw Paciorkiewicz of Poland and manufactured by Gesellschaft für Apparte-und Maschinenbau GmbH from 1908. Three row keyboard was used for letter selection but the linkage was clumsy and the company is said to have made only a few machines, although an example previously in the author's possession had a serial number over 2,000. Originally briefly called Merkur, manufactured by Maschinenfabrik Merkur GmbH, then Phönix and finally Sekretär. Two very similar models, distinguished mainly by the position of the space key, were produced.

A rare machine, so think thousands.

PICHT

Oskar Picht, who spent his life with the blind, was responsible for some excellent machines which were manufactured in Germany in considerable numbers from 1899 onwards. His instruments included one which used six keys and a spacer to emboss braille cells, excellent in design and construction, as well as others using a type wheel operated by a sliding indicator through which the index finger could feel the embossed letters beneath—this design was also offered as a conventional machine for the sighted. Models permitting the blind to type for the sighted were introduced from 1907, shorthand versions using paper tape two years later. First manufactured in Berlin by Wilhelm Ruppert, followed in 1902 by Schreib-und Nähmaschinenfabrik Wernicke, Edelmann & Co., in 1903, by Alois Herde & Wilhelm Wendt and in 1919 by Bruno Herde & Friedrich Wendt.

A pleasant machine, so expect to be out of pocket to the tune of $500-1,000 for a good example. **(214, 215 [c])**

PIONIER

Oskar Picht developed this type wheel machine from his models for the blind by substituting a conventional linear index and indicator. Roller inking. The Berlin firm of Herde and Wendt produced it from 1910.

Price-wise, comparable to the above…only rarer.

PITTSBURG

Front stroke visible with four row keyboard of distinctive elongated profile first produced as the Daugherty by the Daugherty

Typewriter Company and from 1898 as the Pittsburg made by Pittsburg Typewriter Co. of Kittanning, Pennsylvania. Several similar models culminating in the redesigned Model 12 of more conventional format, introduced in 1911, was produced in small numbers before and after the War but failed to save the company from the receiver and the factory finally closed down in 1921. Early models featured interchangeable type bar and keyboard assemblies removable as a single unit. Also marketed under the following names: American, Broadway Standard, Decker-Beachler, Fort Pitt, Reliance, Reliance Premier, Reliance Visible, Schilling, Shilling Brothers, and Wall Street Standard.

Delightful machines, and good value in the $1,000-1,500 bracket. **(216, 217)**

PNEUMATIC

Marshall A. Weir of England was responsible for the first typewriter to use compressed air, patented in 1891 and apparently manufactured, although details are sparse. The design featured thrust-action type bars connected by tubing to a keyboard consisting of three rows of rubber balls, compression of which forced enough air through the tubes to propel the type bars to the platen. Irregularity of impression, slowness of operation and the fatigue obviously induced by this design were all responsible for its swift demise. A model with four row keyboard was also devised.

Difficult to price. Have any survived?

POCKET (1)

One of the smallest machines ever made, this device fits comfortably into the palm of a hand. The knob which turns the enamelled circular index is pressed down for printing. Roller inking. It was first manufactured in England in 1887 and marketed by a variety of outfits which include Dobson and Wynn, the Miniature Pocket Typewriter Co., and the Pocket Typewriter Co. Ltd, 265 Swan Arcade, Bradford—records show this company as occupying that address from 1891 to 1893. Some were labelled Miniature Pocket Typewriter and an example bearing this name, with additional characters and called 'Polyglotte,' was marketed by Bonnet et Cie, Paris—the original receipt, which has survived, is dated 1893. Remarkably enough, extant examples bear serial numbers which run well into four figures.

231. Salter Standard 6

This is a tiny, primitive instrument but rare, and at $2,000 or so, pound-for-pound it is probably the most expensive typewriter in the world.

POCKET (2)

A slightly different circular index machine which was only minimally less primitive was manufactured in the US by the Pocket Typewriter Co. of Rockford. The concept was similar except that the machine ran along a rack clamped over the paper so that at least typing in straight lines could be assured. Date was 1894 and several similar models were produced.

Comparable in price to the above.

POLYGRAPH

This machine marked the entry into the typewriter market of the important Leipzig manufacturer of disc-playing music boxes Polyphon Musikwerke A. G. which had been successfully making their product since 1886. Paul Riessner of Leipzig invented the down stroke design with the curved two row keyboard of Hammond inspiration in 1903; two years later, a three row universal keyboard model was introduced. Production ceased in 1909, although an abortive post-war effort at re-introducing the machine was made in 1919.

Relatively rare, hence up to $4,000-5,000 or so. **(86)**

POLYTYPE

Several models of a multiple type wheel machine for typing up to four originals at the same time, this Italian invention was the work of Felice Molinari of Milano and dated from 1895. The local firm of Prinetti e Stucchi manufactured it in limited numbers. Type wheels, up to four in number, were fitted on a common horizontal shaft above a flat paper table in a frame permitting horizontal and vertical movement. Some models used keyboards, others were index and indicator machines.

None has survived, as far as the author knows, although some northern Italian basement may still conceal one somewhere.

PORTO-RITE

Conventional portable believed to have been made for the mail order trade by Remington in the 1930s.

Low three figures should do it.

POSTAL

Popular type wheel machine highly regarded in its day, the Postal was manufactured in New York by the Postal Typewriter Co. after patents filed by William P. Quentell and Franklin Judge

232. Salter Standard 7

233. Salter Standard 9 (Courtesy of Phillips, London)

in 1903. Three row keyboard, ribbon. Several models, all virtually identical. Contemporary reports claimed sales figures which may well have been exaggerated: 2,000 retailers in the United States alone were said to be selling the machine, as well as agents throughout the main European countries, but serial numbers of surviving examples previously in the author's possession range up to 25,000-odd for a model no. 5, which averages very few machines indeed per agent per annum.

Not particularly rare, but a pleasant machine. Expect prices to approach the four figure bracket. **(219)**

PROTOS (1)

Thrust-action machine of Empire/Adler inspiration manufactured from 1922 to 1925 by Zimmer, Zinke & Co., later simply Zimmer & Co., of Frankfurt. A smaller version patterned after the Klein-Adler and called Protos Kleinschreibmaschine was introduced in 1924.

A few hundred dollars ought to be plenty.

PROTOS (2)

Conventional front stroke uprights better known as Mercedes were said to have been retailed in England under the Protos label, these machines of course bearing no similarity to the above mentioned thrust-action version.

Correspondingly less desirable and thus cheaper.

PUNCTOGRAPH

Machine for the blind dating from 1885 consisting of a wooden frame to which paper was clamped and over which travelled the embossing cell with radial keys.

A couple of hundred ought to be enough for this Stainsby-Wayne clone.

RAPID

Bernard Granville of Chicago can claim the credit for the world's first thrust-action machine which he patented in 1888. Type bars were arranged in four horizontal rows radiating outwards from the printing point and passed through two guide plates on their way to the platen. Despite its historic importance the machine was not a commercial success and production passed through several hands before it was abandoned: Western Rapid Type-Writer Co. of Findlay, Ohio, was replaced by Mead Phillips and Granville of Dayton, Ohio, and A. W. Gump & Co. of the same city is also listed, although possibly as Receiver. Prior to that, the machine had been completely redesigned and labelled New Rapid but without improving the fortunes of the manufacturer.

May not look like all that much but it is rare, so brace yourself for five figures, to avoid cardiac arrest.

234. Sampo (Courtesy of Bernard Williams)

REGINA

Conventional upright manufactured in Germany from 1904 by Schilling & Kramer. Numerous similar models until production ceased in 1934. The machine was a development of the earlier Germania.

Zzzzzzz.....If you are lobotomized from boredom, you might even pay up to forty or fifty bucks, or so.

RELIABLE

Conventional German upright introduced in 1921 by Reliable Schreibmaschinen GmbH of Nürnberg. Also sold under the names Belka, Drugj, Heroine, Libelle, Liga, Mafra, Progress, Saxonia, and Thuringia.

Not to be confused with the Reliance (Pittsburg Visible). If you have shelled out more than a low two-figure sum for a Reliable, you've been turned over.

REMINGTON

E. Remington and Sons of Ilion, New York, had already begun to diversify away from arms manufacture after the end of the Civil War when the entrepreneurial James Densmore thrust the Type Writer at them. Having sunk considerable sums of his own money—some $13,000, in fact—into the project which amateur inventors Christopher Latham Sholes, Carlos Glidden, and Samuel Soulé had been knocking together in Milwaukee, Densmore was anxious to see a return on his investment which their own abortive manufacturing efforts was unlikely to produce. Enlisting the aid of a high-pressure salesman called George Washington Yost and toting the best machine they had so far managed to build, he succeeded in concluding a deal with Remington on March 1, 1873, for the manufacture of 1,000 units with an option for 24,000 more.

Remington was already manufacturing sewing machines **(62)** and placed the project in the hands of their experts in that department who, predictably enough, fitted it to their existing sewing machine tables complete with foot-operated carriage return treadle **(61)**. Production began in September 1873 and the first machines left the factory early the following year, bearing the title 'The Sholes and Glidden Type Writer...Patented...Manufactured by E. Remington and Sons, Ilion, N.Y.'

The machine proved a technical and commercial failure and numerous modifications were made until 1878 by which time the machine had been progressively redesigned to become, after a number of name changes, the famous Remington Model 2. There was no machine actually labelled a Remington No. 1—the designation originated years later when it became necessary to distinguish the first models from the later ones—and all models of earlier vintage than the Remington Model 2 were simply called Type Writers (later Type-Writers, then type-writers and finally typewriters).

1878 was the turning point, and the success of the Remington Model 2 proved that the company had got it right at last. The old Sholes and Glidden sewing machine treadle and table were gone, as were all the covers over the platen, the keyboard, the front and the rear. Upper and lower case had been added by one of their employees called Byron A. Brooks who patented this improvement in 1875, (he later parted company with Remington and manufactured the machine to which he gave his name). The large carriage return lever was removed from the side as was the accompanying housing, and in general the machine assumed a format that was to set the pattern which uprights for decades were to emulate. It remained however an up stroke machine and typing was 'blind,' i.e. not visible to the typist unless the carriage was raised. Remington persisted with this anachronistic feature until 1908, long after competitors were already offering fully visible typing (*'it should be no more necessary for a typist to see the typing,'* they argued, *'than it is for a pianist to see the keyboard'*...but fewer and fewer people were convinced, after trying the visible alternative).

Retailing the machine was entrusted to a succession of agents until 1882, when a former employee called Clarence Walker Seamans together with William Ozmun Wyckoff and Henry H. Benedict formed the partnership which controlled sales world-wide. It was thanks largely to their dynamic policies that the make imposed itself to such an extent upon world markets.

235. Saturn

The successive name changes are as follows:
Sholes and Glidden Type Writer
Improved Type Writer
Perfected Type Writer
Remington.

Despite the appearance of the Model Two in 1878, Remington were by no means convinced that upper and lower case machines were there to stay, and they confidently predicted that they would continue to sell ten upper case only machines for every one offering both upper and lower case. Market forces however quickly proved them wrong, but initially, at least, machines with upper case only continued to be offered side-by-side with those featuring upper and lower cases.

Model Three was similar to Model Two, but with wide carriage and extra characters. Model Four was upper case only, offered principally for typing telegrams received by morse code. Model Five with few notable improvements was introduced in 1888, Model Six in 1894 and the enormously popular Model Seven, which is still to be found today in such numbers, in 1896. Twelve years were to pass before the company finally decided to bow to the inevitable and abandon the up stroke design in

237. Sholes & Gliden, with treadle carriage return (Courtesy of Tom Fitzgerald)

favour of a fully visible front stroke Model Ten in 1908. Subsequent models did not depart from this format.

However, some other notable models based on different principles were introduced at various times and received mixed receptions. A three row double shift machine called Remington Junior was launched in 1914; six years later saw the arrival of the oblique front stroke Remington Portable in which the type bars could be lowered by means of a knob on the right of the machine to make it more compact for carrying, and 1928 saw the brief resurrection of the Blickensderfer Model 5 under the name Rem-Blick and Baby-Rem.

Meanwhile, the factory had already deviated from producing strictly front stroke machines after it had bought up the Noiseless Typewriter Co. in 1924 and renamed the thrust action product Remington Noiseless, retaining the same mechanical principles on the four row keyboard Model 6, introduced in 1925, and on subsequent models.

Production of the conventional upright format, in common with other manufacturers, continued until after the Second World War, when stylistic changes common to the times were first introduced. Electrics appeared from the early 1950s, although, again in line with market trends, an abortive attempt at electrifying a manual upright was marketed briefly in 1925. Most surviving up stroke Remingtons are Model 7, of which there are many in circulation so condition is pivotal in determining prices even up as high as the $500-1,000 bracket. Expect to pay marginally more for Models 5 & 6 than for Models 2 & 7. Conventional upright Remingtons are ubiquitous, hence cheap, and the same applies to portables. Rem-Blicks and

236. Sholes & Glidden, with treadle carriage return[41]

238. Sholes & Glidden, on sewing machine table top

Baby-Rems: cf. Blickensderfers. **(89, 220, 221, 222, 223, 224)**

REM-SHO

As the name implies, this machine was the outcome of the collaboration of Messrs. Remington and Sholes, but not of the original typewriter pioneers. Their sons Franklin Remington and Zalmon Sholes formed the Remington-Sholes Typewriter Co. and began manufacture in 1896 of an up stroke machine with four row keyboard which had been patented two years earlier largely on the strength of an original shift mechanism which displaced the entire basket of type bars rather than the carriage. Legal battles over the use of the name Remington on a typewriter (even though the family itself had long since divested itself of its typewriter interests) plagued the company during the greater part of its life-time: when Franklin and Zalmon lost the early rounds, the name of the machine was changed to Fay-Sholes and the manufacturing company to Fay-Sholes Typewriter Co. (Charles Norman Fay was its president) but legal appeals went all the way to the Supreme Court, which reversed the judgment, and the typewriter was once again marketed as the Rem-Sho (also Remington-Sholes and Remington-Fay-Sholes).

Use the weight test: early examples of the Rem-Sho are highly decorative and cast in bronze, and are heavy to lift. Later the castings were merely bronzed and the machine is much lighter in weight.

The name on the paper table was stamped out of the metal, as on a stencil.

Model 10 of conventional upright profile offering visible typing was introduced in 1908 but it did little for the fortunes of the company which went into receivership the following year, its assets sold to the large French manufacturer Japy Frères. Understandably enough, bronze (not *bronzed*) models fetch the best prices in the lower half of the four-figure spectrum. Others are correspondingly less desirable and hence cheaper. **(225)**

RHEINMETALL

Conventional upright manufactured in Germany in numerous similar models between the Wars by Rheinische Metallwaren- und Maschinenfabrik, later Rheinmetall-Borsig A. G. First introduced in 1920. Also labelled Metal, Rheinita, and Rheinita Record. A conventional portable of low profile was introduced in 1931.

Germans are on record as having been tempted to spend up to three figures for one; other nationalities generally walk straight past them (the machines, that is).

RICO

Pleasant toy typewriter with indicator selecting letters from a rectangular index. Prices up to $500 or even more, reflecting relative interest value. **(227)**

ROCHESTER

Wellington P. Kidder, better remembered for the thrust-action machine to which he gave his Christian name (and for its various derivatives) was responsible for this small portable with three row keyboard and double shift. Also typing by thrust-action, the machine was manufactured in 1923 by Rochester Industries Inc. of Rochester, New York, after it acquired the assets of the Leggatt Portable Typewriter Corporation which was to have produced a machine with similar specifications. The machine was originally intended to be marketed as the Midget.

Probably four figures, but the question is how high.

ROFA

Down stroke type bar machine with curved three row keyboard and double shift. Inking by roller, and this combination of type bars and roller inking, while by no means unique, is nevertheless rare and is all the more unusual given the fact that the machine was introduced as late as 1921. Robert Fahig GmbH which had made the Faktotum before the War was a manufacturer. The 1923 Model 4 had a straight three row keyboard. Sold in Holland as the Correspondent.

$500 is about the ceiling, which is good value. **(11)**

ROYAL

This major name first appeared on the market in 1906 with the introduction of the so-called 'flat-bed' model of low profile, the invention of E. B. Hess of New York who had already devised an unsuccessful nine-bar machine typing eighty-one characters—only one of the 140-odd patents of which this prolific inventor could boast.

Front stroke, type bar machine manufactured by Royal Typewriter Co. of New York. The flat-bed design continued through minor model changes for several years until the familiar conventional upright profile was adopted on the Model 10 in 1914. Aggressive sales promotion included dropping machines out of airplanes to prove their solid construction.

239. Sholes & Glidden No. 33: typing sample (approx. 8 pts)

THE SHOLES AND GLIDDEN TYPE WRITER NO. 33 IS THE OLDEST SURVIVING EXAMPLE OF THIS MACHINE STILL LEFT IN PRIVATE HANDS.

Production of similar uprights in cosmetically different models continued into the 1950s, in line with general market trends. Electrics were also added to the range in the 1950s and noiseless and portable models were introduced at various times to complete the range and compete on all levels.

By the 1950s, Royal was calling itself the 'World's Largest Manufacturer of Typewriters.'

Not to be confused with Royal Bar-Locks, which have nothing to do with the Royal machines.

The early models make up to a few hundred dollars, particularly the pleasant glass-panelled ones which are a good value for the money. Later models jostle for vantage points on the two-figure ladder. **(97, 228, 229, 230)**

ST. DUNSTAN'S BRAILLE WRITER

Machine for the blind of standard specifications produced by the well-known organization of that name.

Low three figures, on a sunny afternoon.

SALTER

George Salter and Co. of West Bromwich, best remembered as manufacturers of scales and weighing machines, entered the typewriter market with a fine down stroke type bar instrument patented in 1892 by James Samuel Foley and John Henry Birch. Type bars were disposed vertically in an arc, from the earliest examples up to the introduction in 1913 of an oblique front stroke model offering improved visibility of typing. Three row curved keyboard on early models, replaced by a straight three row from 1900 with the introduction of Model 6, the keyboard virtually unchanged to 1913 when it was superseded by a conventional straight four row keyboard.

The first model, labelled simply The Salter Typewriter, was replaced by the Improved No. 5 at around serial no. 2000 or so and the specifications were finally stabilized. To that point, several modifications had been introduced in quick succession, all of them relating to inking mechanisms. The 1892 patent itself specified a mechanism which was actually never manufactured, in which the type faced backwards, resting against an inking pad and performing a half-revolution as the type bars descended to the printing point. By the time the machines were first produced, however, the type bars faced forwards and inked against a roller. This was undoubtedly a simpler mechanical construction, but the inking system was primitive and unsatisfactory and was soon replaced by a narrow ribbon. These successive changes appear to have been introduced on serial numbers only several hundred units apart, although attempts at a more precise and positive identification using serial numbers has not proved conclusive. Of early Salters previously in the author's collection, serial no. 1071

240. Sholes & Glidden No. 2760, with side handle

241. Sholes Visible (Courtesy of Bernard Williams)

had roller inking, no. 1725 had no inking mechanism whatever but was probably roller since no trace of a ribbon mechanism existed on the machine, while no. 878 sported a ribbon, which leads one to suspect that it was originally a roller model, later modified to ribbon.

All the above were labelled The Salter Typewriter. The Improved No. 5—the improvements limited to such details as ribbon width and advance, paper guides and carrying cases—was replaced by No. 6 in 1900, with serial numbers of machines previously in the author's collection up to 7698, then Improved No. 6 (the earliest example known to the author was his serial number 7803), then No. 7 in 1907, and Model 9 in 1908. The oblique front stroke model with four row keyboard appeared in 1913.

Also marketed under the names Rapide, Perfect, and Royal Express.

Pre-ribbon models are likely to fetch five figures. Expect No. 5 to make anywhere up to $5,000 or so, later models progressively less than $1,000. The oblique front stroke Salter Visible in the $2,000-4,000 range: grey examples make more than black. **(22, 88, 91, 92, 117 [I], 231, 232, 233)**

SAMPO

500 machines were said to have been built to the specifications of a Dr. Raphael Herzberg of Helsingfors by the Swedish company Husqvarna Vapenfabriks A. B. but were apparently unsatisfactory and nothing further is known of this first Swedish machine except that it was also called Finnland possibly with the market in that country in mind. Date was 1894.

Brace yourself for a price tag of up to five figures. If you are offered one for less, pay up and run for it. **(234)**

SAMTICO

Samuel Timings contributed the necessary letters of his name to this primitive machine which he and John Burgess patented in 1897 and which the Timings Burgess Co. of London produced two years later. Three alternative designs were protected in the patent specifications: separate characters held in place by a rubber band, a circular index with the letters stamped out as on a stencil, inking by pad, and another circular index design similar to the Simplex.

Prices comparable to Simplex.

242. Sholes Visible (Courtesy of Tom Fitzgerald)

SATURN

Unusual design which featured a rectangular index of seventy-two upper and lower case characters arranged in nine columns of eight characters each. Each column had its own printing key which was pressed for typing, the letter in the particular column having been first selected by means of a grip on the left attached to a cord which ran the length of the index. This procedure selected the desired letter on a plate containing rows of sliding type which was located beneath the platen. The typing was blind, performed by pressing the column key. Ribbon inking.

The machine was actually manufactured, but understandably very few were made. F. Meyer-Teuber invented it in 1897 and Feinmaschinenwerk E. Stauder of Meilen, Switzerland, manufactured it two years later.

Also referred to as Stauder, after the manufacturer, although whether the instrument was actually marketed under that name is uncertain.

Four figures, for sure. **(235)**

SCHADE

It is indeed a pity that the radial type plunger machine of this name was not made some thirty years earlier, for it might have earned itself a different place in history. Rudolf Schade of Berlin was responsible for the device (one hesitates to call him the inventor) and a few examples were made in 1896. The machine was virtually identical to Pastor Malling Hansen's Skrivekugle except that it was fitted with a clamp for attaching to a table-top. Plungers were arranged with the space key in the centre, surrounded by lower case keys, then upper case and finally, towards the periphery, figures and signs. Ribbon. The inventor maintained to the bitter end that he was not guilty of plagiarism, which you may believe if you wish.

Somewhere in the five figures, is the best educated guess.

SECOR

Conventional front stroke upright invented in 1905 by Jerome B. Secor and manufactured by Secor Typewriter Co. of Derby, Connecticut, in the former Williams factory. Some 7,000 or so were built.

Rare, but potentially of real interest only to Derby (Connecticut) Historical Preservation Societies and their members, although the Williams connection may help to enhance the Secor's value well into three figures.

SEIFRIED

Typewriters and shorthand machines for the blind, manufactured in the United States in the 1890s. One of the machines, called Midget Writer, which had three keys and a spacer, embossed braille cells on paper inserted into the rear of the instrument between two feed rollers. A stenographic version of this machine was introduced in 1900.

A couple of hundred should be plenty.

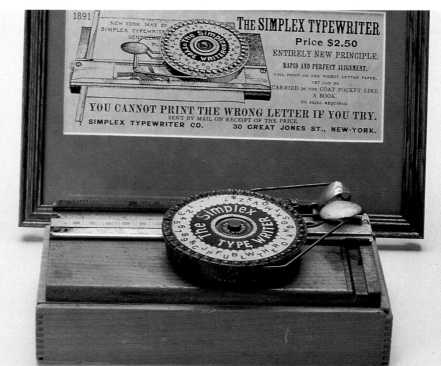

243. Simplex (Courtesy of This Olde Office, Cathedral City, California)

244. Smith Premier No. 1, detail
(Courtesy of Tom Fitzgerald)

SENATOR

Conventional German upright produced in limited numbers from 1922 by Falcon-Werke A. G. of Darmstadt.

See price comment under Secor, above, but substitute Darmstadt for Derby.

SENTA

Conventional front stroke machine with three row keyboard and double shift produced in 1913 by the German sewing-machine manufacturer Frister and Rossmann A. G. which had already entered the typewriter market by producing the Caligraph and the Moya under licence. Four row model introduced in 1926. Production ceased in 1930. Also sold under the names Presto and Balkan.

$100-150 should be ample.

SHIMER

Up stroke machine with a four row keyboard patented by Elmer S. Shimer of Milton, Pennysivania, in 1891. The machine bore more than a passing similarity to early Remingtons but limited manufacture was nevertheless undertaken.

Not intrinsically exciting, but rare…which probably means four figures, does it not?

SHOLES AND GLIDDEN

Christopher Latham Sholes, a Milwaukee printer and amateur inventor, was busy designing a numbering machine when it was suggested to him that the device could be made to print letters as well as numerals. Together with fellow inventors Carlos Glidden and Samuel Soulé and with technical assistance from a machinist called Schwalbach they knocked together a crude model of a typewriter built into a converted kitchen table. It was September 1867 and within a month of completion, the patent application was filed.

The table had a circular hole in the top, beneath which hung a circle of type bars connected by means of wires to key levers on one side. On top of the table and over the hole was a frame connected to a weight-driven escapement. The frame held a sheet of thin paper with a ribbon above it, the type face striking paper against ribbon.

Work on improving this crude model continued but was hampered by lack of funds and it was at this point that a dynamic albeit unpleasant investor in the project was found. James Densmore bought himself a 25% share sight unseen by reimbursing Sholes, Glidden, and Soulé the $600 of their own money which they had so far expended, but when he eventually saw the machine some months later, Densmore demanded improvements and a smaller and more compact machine was produced. This now had a piano keyboard in place of the levers, but the up stroke type bars were retained as was the flat paper carriage; Densmore filed the patent application on this model on 1st May, 1868.

Anxious to see a return on an investment he could ill afford, Densmore imprudently began limited manufacture of this so-called improved model, but fifteen defective machines and $1,000 later, it was back to the drawing board for Mr. Sholes. More of Densmore's money went into the project, and some twenty-five or thirty prototypes (possibly even more), each embodying some improvement or other, were built before the machine was once again pronounced perfect. By now, the instrument was able to boast a cylindrical platen in place of the flat paper carrier, although typing was around and not along the platen, and the piano keys were replaced by neat brass buttons. It was essentially this design which the partners patented in 1871; the improvements, such as they were, did

245. Smith Premier 60
(Courtesy of John P. Lewis Sr., Albuquerque, New Mexico)

not succeed in ironing out the bugs, however, and the partners began selling off their share holdings for whatever they could get.

All except Densmore, that is, who never lost faith. Together with a smooth talker called George Washington Yost (who was later to produce the machine to which he gave his name), he succeeded in interesting E. Remington and Sons of Ilion, New York, in the project and the result was a contract signed on March 1st, 1873, for the manufacture of 1,000 units, with a further 24,000 option. Densmore had to find the $10,000 advance, but the worst of his problems appeared to be over.

Remington was ready to begin manufacture in September 1873, having entrusted the prototype to their sewing machine division, which predictably enough, mounted the typewriter on their sewing machine table complete with foot-operated carriage return treadle. It was fully-enclosed, with covers front and rear, with a lid over the platen and a flap over the keyboard. Embellished and decorated with scrolls, flowers, and pictures, and titled 'Sholes and Glidden Type Writer, manufactured by E. Remington and Sons, Ilion, New York' it was offered to an unenthusiastic public by Densmore and Yost for the not inconsiderable sum of $125 a machine.

Debts were piling up as fast as unsold machines and even when sales were made and money came in, there were so many claims on it that it is small wonder the venture foundered. Creditors and minor shareholders on all sides were clamouring for money, and there were royalties to be paid as well. Densmore had lost control, and Yost was taking over.

Ultimately, in desperation, they concluded a new agreement with the ailing Remington company in November 1875 in which their interest was reduced to a fee per machine for the use of their patents. The retailing of the machine passed from Densmore, Yost and Company to Locke, Yost and Bates, and eventually to Fairbanks and Company all in swift succession, with bankruptcy staring them all in the face.

But the turning point had been reached. Three men, who had already been concerned with the venture for some years in one capacity or another—Wyckoff, Seamans and Benedict—formed the dynamic sales organization which first promoted and sold the Type Writer and successive Remington models and which eventually, in 1886, bought up the Remington typewriter division altogether, name and all.

As for the machine itself, which was at the heart of all this frantic activity, the Sholes and Glidden went through several progressive changes. It began its manufactured life fitted to a sewing machine table with a foot-operated carriage return treadle. It had a large diameter platen, which could be protected by a cover hinged at the back. It had further covers front and rear concealing the maze of wires from inquisitive fingers, and it had yet another cover over the keyboard. An arm swivelled on the left for holding notes and text being typed. It had a keyboard of bold upper case letters, and of course it printed only in upper case. Key levers were wood, as was the space bar. At the top of the keyboard was the wooden bar on which the serial number was impressed.

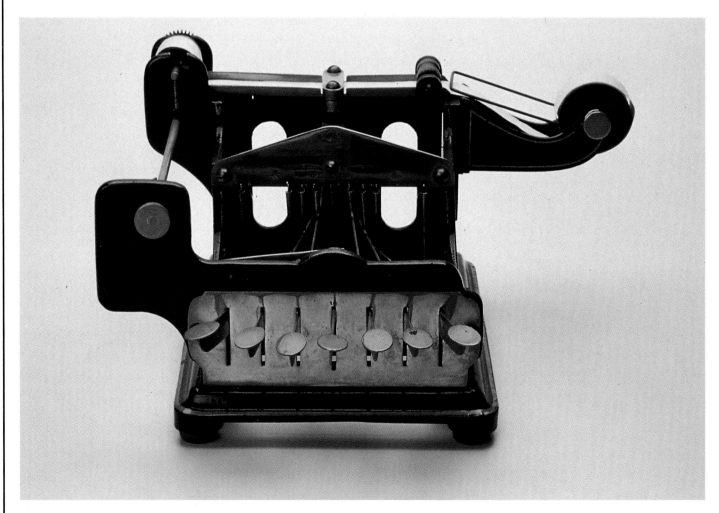

246. Stainsby Wayne [typewriter model] (Courtesy of This Olde Office, Cathedral City, California)

It was not long, however, before some obvious changes were introduced. The treadle mechanism was discarded and replaced by a handle on the right-hand side—this was the most glaring and obvious 'improvement,' although whether pressing down a large handle was easier than pushing on a treadle is debatable, and this must have occurred to the manufacturers as well, for they soon abandoned the side handle in favour of the smaller lever on the carriage itself. Meanwhile, it was becoming obvious that all the covers were not really required and these were also discarded and the 'open-frame' machine substituted. And finally, a shift mechanism was added providing upper and lower case.

Naturally enough, name changes accompanied these modifications and The Type Writer of 1874 was soon being offered in different models. By 1877, before it was being called a Remington, it had already been known as the Improved Type Writer and then Perfected Type Writer. Each of these versions was offered in four and possibly five models, including one typing both upper case and lower case which was being sold simultaneously with the upper case only models, it being by no means certain, at the time, which format would ultimately prevail.

The turning point in the Type Writer's fortunes, however, may be attributed directly to the introduction of the Model 2 which proved to be one of the most successful machines ever made. Its success was not immediate, however, nor was it mechanically perfect from the start, but by the time Remington put their name to it in the early 1880s calling it the Remington No. 2 its success was assured.

According to an article in an 1886 magazine, some 3,000 Sholes and Gliddens were said to have been built.[46] There were almost certainly more, for serial numbers of surviving machines have been traced right through into the 4,000s, and examples previously in the author's collection alone range from serial no. 33 to no. 3879, while the London Science Museum Collection has one with the serial number 4978.

Altogether, upwards of 100 examples are known to have survived to the present time, many of them up-dated by the factory to later specifications and their serial numbers stamped with the prefix 'A.' It would appear that well over 1,000 were originally treadle machines, and a further 2,000 or more were originally fitted with the side carriage-return handle. The small return lever fitted to the carriage itself was a late modification and was retained on the Remington Model 2, and later models.

A detailed study of surviving S & G's reveals so many differences from one machine to another that despite its limited mass production, manufacture was still very much a one-off job. Certainly in terms of decor this would appear to be the case—very few are alike. We ought not to be as surprised by this as we are. I recall that I once needed to visit the Scammell commercial vehicle factory in the north of England to discuss some details relating to some Highwayman trucks I had received. To my amazement, they opened up a hand-written ledger to the vehicles in question and said 'Oh yes, Jack built

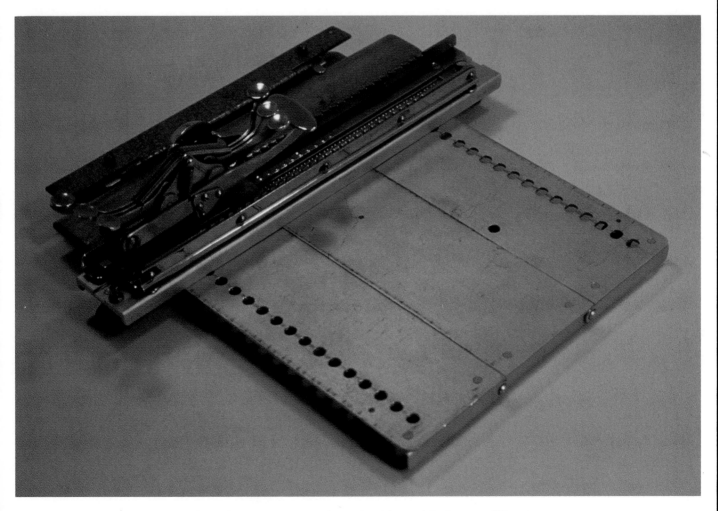

247. Stainsby Wayne [shorthand/telegraphic model] (Courtesy of This Olde Office, Cathedral City, California)

those' and a few minutes later, there was Jack saying 'Yes, I remember those trucks. We had a problem with one of the rear spring hangers on one of them…' That was in the early 60s…and not the 1860s, either. Little wonder, then, that early Sholes and Glidden machines had such a diversity in minor specifications.

It is worth further note that some Sholes and Glidden machines have recently come to light with obviously modified specifications, but without the 'A' prefix to the serial number. The explanation has been offered that perhaps the factory simply forgot to stamp the letter before the serial number, after carrying out the modification. This is simply not plausible and is almost certainly incorrect, for the workers responsible for the modifications would be no more likely to forget that

part of the operation than any other. What is far more likely is that the modifications were carried out by some private workshop using parts supplied by the factory, possibly in the case of machines shipped overseas.

Not as rare a machine as one might think, but no collector can sleep until he has one, which of course tends to inflate prices. A complete and un-modified treadle machine would fetch the highest amount. Price considerations include a premium for low serial numbers, originality and completeness of specifications including side panels, top cover, and original decorations, etc. Side handle examples tend to fetch higher prices than modified earlier ones, or unmodified later ones.

Count on five figures, of course: the question is only how high. **(Rear cover, 1, 12, 51, 61, 62, 236, 237, 238, 239, 240)**

248. Stainsby Wayne shorthand/telegraphic model

249. Standard Folding (Courtesy of This Olde Office, Cathedral City, California)

SHOLES VISIBLE

Probably the last typewriter designed by Christopher Latham Sholes and patented by him in 1891, this front stroke type bar machine with four row keyboard was manufactured by the inventor's son Louis first with the Meiselbach Typewriter Co. in 1901 and again later, in 1909, with the C. Latham Sholes Typewriter Co. Both attempts were unsuccessful and were soon abandoned.

Type bars in two parallel rows at right angles to the platen provided the basis for the unique feature of this machine. When a key was pressed, the corresponding type bar first moved into the space between these two rows and then onto the platen, returning by the same route to its place of rest. It was guided all the way, in an attempt at overcoming alignment problems. Typing was visible.

Also said to have been sold as the Bonita Ball Bearing and the Meiselbach.

Around $3,000 or more, for a good one. **(102, 241, 242)**

SHORTWRITER

Shortwriter Co. of Chicago produced a stenographic machine using twenty-four keys arranged in two groups of ten, one for each hand, plus four vowel keys for the thumbs. Introduced in 1914, the machine printed on paper tape and was sold in considerable numbers.

Several hundred should be enough.

SIEMAG

Conventional German upright introduced in 1949 by Siegener Maschinenbau A. G.

If someone offers you one, offer to take it off his hands without charging him!

SIMPLEX (1)

Tinplate circular index machine mass produced in numerous models by Simplex Typewriter Co. of New York after a patent granted in 1892 to A. M. English of the same city. This inexpensive machine was sold principally as a toy but became more sophisticated as time went by. Early models distinguishable by numbers or letters of the alphabet or both—Model 1, 1A, A, B, E etc. The principle remained the same, and differences were largely cosmetic. Some were launched for specific occasions, eg. 'Simplex Souvenir 1903.'

Many later patents followed, most of which were granted to Philip Becker (to whom the 1892 patent was assigned) and to W. Thompson, both of New York.

Marketed under the names Practical, Baby Practical, Baby Simplex, Little Giant, etc. In England it was also called Gladstone, and in France Eureka **(159 [top c])**.

Very many slightly different models make this an esoteric field for aficionados and cognoscenti. By rights, a two or low three figure sum ought to be sufficient to conclude a deal but it is unlikely to be enough, so be prepared for higher. **(159 [centre l, lower r], 243, 268 [c])**

SIMPLEX (2)

A primitive typewriter identical to the 'Dollar' **(159 [lower l])** with its vertical type wheel, and labelled Simplex Model 1 was manufactured by Robert Ingersoll of New York, manufacturer of Dollar watches etc.

'A very paltry invention' but the price tag for paltry inventions can be hefty these days.[62] Prices up to four figures have been recorded but ought not to exceed the middle hundreds.

SIMPLEX (3)

The Smith Premier Model No. 10 stripped of such accessories as tabulator, two-colour ribbon etc. in an effort to reduce price was sold in Germany prior to the First World War under the name Simplex.

Not as common as the straight Smith Premier model but the price is likely to be roughly the same.

SKRIVEKUGLE

One of the rarest and most desirable machines which collectors all over the world dream about but few possess, the Skrivekugle as it was known in Danish, or Writing Ball as it has come to be called in English, was the work of Pastor Hans Johan Rasmus Malling Hansen, first a teacher and eventually a director of the Copenhagen Deaf and Dumb Institute. It was in that capacity that he began designing his writing machine, the first model of which was completed in 1865, although the oldest surviving model, and the one which was to provide the prototype for eventual manufacture, was produced two years later.

The characteristic feature of the machine, from the start, was the mechanical principle on which it was built. Of radial type plunger design, the keyboard was essentially the sector of a sphere with type bars projecting radially outwards, from a common printing point underneath, to the neat brass key-tops above. The action was therefore direct and free from the cumbersome linkages which plagued rival inventors. This principle was not original to Malling Hansen but his was its first successful commercial application.

250. Sterling (Courtesy of Bernard Williams)

Other features of the design included a cylindrical platen operated by clockwork with an electromagnetic escapement common to early printing telegraphs. Carbon paper for the impression, with the whole apparatus enclosed in a wooden box so that only the keyboard (or key-ball) protruded above the hinged lid into which it was fitted, and which had to be raised to reveal the typing.

That 1867 machine, placed in the hands of the Jurgens Mekaniske Establissement, was subjected to developmental work lasting more than two years before it was ready for series manufacture. The 1870 product, a massive 75 kgs of precision engineering, retained the keyboard which now sported fifty-two plungers and retained the principles of the electro-

251. Sugimoto: prototype (Courtesy of Yamada Hisao)

magnetically controlled carriage escapement, but was completely open, mounted on a massive brass frame with a flat paper table, or a cylindrical platen as an optional extra.

It was offered for sale in London in 1872 in that form for £100, protected by an 1870 British patent which specifies both the flat paper table and the cylindrical platen, and protects a mechanical escapement as well as the electromagnetic version. It also proposes another fascinating application: if notches were cut directly into the sides of the type plunger, then pressing the plunger could be made to break and close an electrical circuit for simultaneous printing and telegraphic transmission.

けいざい
博物館

出されました。値段は180
円。いまの60～70万円でし
ょうか。

める2,400字の選定、並べ
方だったといいます。

日本タイプライター狭山
製作所に保存されている和
文タイプライター第1号＝
写真＝です。大正6年5
月、「邦文タイプライタ
ー」の商標で本格的に売り

製作者は、岡山県出身の
杉本京太氏（昭和47年没）
で、大正4年に試作機を完
成、さらに改良のうえ発売
されたのがこれです。仕組
みはいまとほとんど一緒。
苦労したのは、文字面に納

しかし、そのタイプライ
ターも昭和53年秋に東芝が
日本語ワープロＪＷ－１０
（630万円）を発表するに
及んで致命的な打撃を受け
ました。日本タイプライタ
ーも一昨年夏、キヤノング
ループ入りしました。

252. Sugimoto: 1931 Asahi newspaper illustration

Further patents followed. One in 1872 protected a modified portable version which featured a rather fanciful scheme for embossing strips which could later be passed through a stationary printer, while a further refinement patented in 1875 protected a curved swinging paper carriage with printing against a central block. This version, again beautifully made, was produced in some numbers, although few have survived, and a final model using a conventional platen was also manufactured.

It was a wonderful machine, infinitely superior to the rival Sholes and Glidden, and it was sold in quite some numbers throughout the world, winning prizes and awards and rave reviews wherever it went. As late as 1886, the open-frame model current at the time 'appears to be the only foreign competitor American type writers have' and while that was not strictly accurate, of course, it indicates nevertheless the respect with which the Writing Ball was treated.[46]

Also briefly manufactured under licence in Austria by Albert von Szabel. Later look-alikes included an 1885 cipher version by Alexis Kohl called Cryptographe and the Schade type plunger machine of 1896.

The Hansen Writing Ball is a machine to die for, and the price one might reasonably expect to pay is commensurate with its desirability. Sotheby's in London sold one in 1989 for the all-time high of approximately $40,000, which was over-the-top, but if you get one for less than $15,000-20,000, consider that you have done well! **(Frontispiece, 4, 43, 58, 59)**

253. Sugimoto: early model Nippon (Courtesy of Yamada Hisao)

SMITH CORONA

Product of the 1926 merger between the L. C. Smith Typewriter Co. and the Corona Typewriter Co. The Smith brothers first rose to fame and fortune with their Smith Premier, and when their company joined Remington and others to form the typewriter trust called Union Typewriter Company in 1893, they occupied prominent positions on the board of that organization. In 1903, however, they resigned *en masse* to form the L. C. Smith & Bros. Typewriter Co. and market the machine of the same name. Visible typing was allegedly the reason for their resignation from the Union board, the Remington dominance and insistence on the 'blind' design being immovable.

L. C. Smith & Corona Typewriters Inc. followed general trends during the 1930s and 40s and succeeded in cornering a significant portion of the market. The name was contracted simply to Smith-Corona, until their merger with Marchant Calculators in the 1950s.

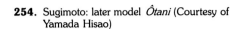

254. Sugimoto: later model *Ôtani* (Courtesy of Yamada Hisao)

255. Sugimoto: portable army model **(a)**, in use **(b)**

If you pay half of what you feel decency demands in all honesty, you have probably paid too much, although really good examples are likely to set you back a low three figure amount.

SMITH PREMIER

Alexander Timothy Brown of New York invented this important machine with patents dating back to 1887 when it was first produced by the Smith Premier Typewriter Co. of Syracuse, New York. Up stroke type bar machine with double keyboard, its distinctive features including a space bar for each hand, a ribbon which moved in a zigzag so as to utilize its entire width (this feature had already appeared on the earlier Brooks) and a circular brush wound by a crank inside the type basket for cleaning the type which was of concave profile to match the curvature of the platen. The type bars were short in length and mounted in wide bearings, all of which helped maintain proper alignment. It was a fine machine and serial numbers reached well into six figures, of which a significant number appears to have survived to the present day.

First model with its decorative panels was replaced by the similar but plain Models 2, 3, and 4 in 1896 (the difference between these models limited to such details as number of keys and length of platen). The up stroke action was eventually replaced in 1908 by the front stroke Model No. 10 which still, however, retained its double keyboard. It was not until 1921 that the name Smith Premier first appeared on a conventional four row front stroke upright which had previously been marketed as the Monarch before these two manufacturers, together with Remington and others, joined forces to form the Union

Typewriter Co. trust. The Smith Premier name continued to be used on machines more widely marketed as Remingtons until the Second World War. Some models of the Remington Noiseless were also briefly marketed as Smith Premiers. A range of carriage widths was offered, including the rare 1923 model 60 with the monster carriage for accounting purposes. **(245)** Smith Premiers were very good machines…which means that of most models many examples have survived, thereby depressing prices. The first model tends to make the highest price, somewhere between $500-1,000; later models correspondingly and considerably less. **(96, 244)**

SMITH VISIBLE

The Yetman machine was briefly marketed under this name after J. L. Smith of Philadelphia bought up the assets of the manufacturer when it went into liquidation in 1909.

Value comparable to the Yetman in the $2,000-3,000 bracket.

SPHINX

Not Egyptian, but Swiss. Conventional front stroke with four row keyboard and recessed laterally-placed ribbon spools. Initially introduced in 1913 by S. A. Sphinx of Fleurier, it was made in very limited numbers until the outbreak of the First World War after which a fresh but abortive effort was made in 1918 to continue production.

Not commonly seen, but does anyone care…apart from the Swiss, perhaps. If it gets up a decent speed down a grand slalom run, it might make three figures.

STAINSBY-WAYNE

Important manufacturer of machines for the blind, both stenographic and braille embossing. The first effort patented in 1899 by Henry Stainsby was a braille machine with six keys and cell spacer which rode on rails over a flat base. Different models followed, including a small upright which perforated paper tape from a roll 200 yds. long, with a warning bell when paper broke or came to an end. One hundred and forty words per minute were claimed. Manufacturer was Alfred Wayne & Co., Handsworth, Birmingham.

Not more than a very low three figure amount should be more than enough to secure it. **(215 [l & r], 246, 247)**

STANDARD-FOLDING

Forerunner of the ubiquitous Corona, this small lightweight portable was patented by Frank S. Rose of New York in 1904 and manufactured first by the Rose Typewriter Co. in 1907

256. Sun (Courtesy of This Olde Office, Cathedral City, California)

replaced by a combination of the numbers 1 to 5 for each hand. An Italian version of this second machine produced by Giulio Crespi of Milan in 1903 and called Stenodattilografica increased the number of keys to twelve by the addition of a zero to the code.

A hundred or two ought to be plenty.

STENOGRAPH

The United States Stenograph Company of St. Louis manufactured a shorthand machine by this name in 1889, patented ten years previously by Miles Bartholomew. It was a successful design using ten keys to produce a code of strokes on paper tape, which on more sophisticated models was advanced by clockwork.

Orthographic niceties were dispensed with, in the system proposed by the manufacturers—the message 'We are never so near playing the fool as when we think ourselves wise' is written as 'W r nvr so nr plang th fl z hn w thnk rslvz wiz.'

Prices might go all the way up the mid-three figures, more than likely.

STÉNOPHILE

Charles Bivort invented this excellent shorthand machine in 1906 and its production continued until the outbreak of the First World War. Each hand operated ten keys, with a space key centrally positioned. Like-sounding consonants (e.g. K, G, C, Q) were allotted the same key, which produced a symbol that looked like a combination of them all. Also offered in Germany in 1913 as the Dictograph.

Barely into the low three figures ought to be about right.

and two years later by the Standard Folding Typewriter Co., both of New York. The design featured a carriage and platen which could be pivoted forward to rest upside-down over the keyboard thereby making it more compact for carrying. Three row keyboard, front stroke with double shift; it became the similar but less handsome Corona in 1912 and the manufacturing company changed its name to Corona Typewriter Co. in 1914. This was only after a troubled few years, particularly 1909, when control of the company changed hands several times and its future appeared precarious.

Nice one, and genuinely good value at prices generally in the $300-500 bracket, or thereabouts. **(249)**

STEARNS

Conventional upright produced by the bicycle manufacturer E. C. Stearns and Co. of Syracuse, New York, from 1908, after three years developmental work. The machine enjoyed considerable success on both sides of the Atlantic and went through several (mainly cosmetic) model changes. Marketed in Germany under the name Guhl, by Guhl and Harbeck, whose earlier products had included such illustrious collectors' items as Hammonia and Kosmopolit.

Prices of up to a few hundred dollars have been paid. **(99)**

STÉNODACTYLE

Frenchman called Jules Lafaurie produced his first model of what was to become a popular shorthand machine in 1899. Combinations of dots on vertical lines were produced by five keys for each hand which could be pressed singly or in combinations. Several models were produced; the original shorthand code was abandoned in 1902 and

257. Taurus (Courtesy of the late Josef Zimmerman)

STÉNO-TÉLÉGRAPHE

A transmitter at one end synchronized with a receiver at the other was at the heart of a shorthand telegraphic invention presented in Paris in 1886. At first, a separate wire was required for each character; later a single line was sufficient. Perforated paper tape similar to that used on player pianos increased the speed of transmission to more than 200 words a minute, it was claimed.

Rare. Value? Anybody's guess, but probably not excessive since interest in this sort of instrument tends to be specialized.

STENOTYPE (1)

A successful design using twenty-four keys, patented by Ward S. Ireland of Dallas in 1911 and produced by the Stenotype Company of Indianapolis and later of Owensboro, Kentucky, and others. Early models printed a special code of long and short lines but this was later changed to Roman characters. Left and right hands each controlled consonant keys and the thumbs the centrally located vowel keys. Ireland was responsible for a similar later design marketed from 1916 under the name National.

Very low three figures.

258. Tip Tip (Courtesy of Bernard Williams)

STÉNOTYPE (2)

This famous French shorthand machine was the invention of Marc Grandjean. Each hand controlled ten keys angled to conform to ergonomic principles and the keys were pressed in chords. It was introduced in 1910 and manufacture continued for decades. In England it was sold under the name Stenowriter. A lovely machine, well made but plentiful so $100 or so ought to be sufficient.

STENOTYPER

Franklin Hardy of Chicago patented a shorthand machine in 1897 which limited the keyboard to three keys per hand plus a centrally located space key, printing a code of dots and dashes on paper from a roll. In Germany it was manufactured under licence by Adlerwerke vorm Heinrich Kleyer, A.G. from 1899. A couple of hundred.

STERLING

This name appears on a second attempt, made in 1910 or 1911, at manufacturing a swinging sector machine originally called Eagle: the two machines were virtually identical.

And roughly identical in price, too. **(250)**

STOEWER

Bernhardt Stoewer A.G. of Stettin (Szczecin) was one of several German sewing machine and bicycle manufacturers who branched out into typewriter production. The front stroke up-

right originally with three row keyboard and double shift designed by Paul Grützmann in 1903 was soon modified to conventional four row specifications. Three row portable model introduced in 1912 was replaced in 1926 by a four row model, which, from 1931, was manufactured in the Rheinmetall works and marketed under their own name. At different times Stoewers were labelled Baka, Barratt, Cito, DS, De Esse, Elite, Lloyd, Stoewer Record, Swift, Swift Record, and Swift Visible. Understandably enough, German collectors seem most likely to want one for prices up to a couple of hundred dollars. **(154)**

SUGIMOTO, KYOTA

Japanese typewriter pioneer, whose 1915 patent became the prototype of the modern machine used in that language until the advent of the word processor. This was presumably a feat of sufficient significance to warrant Mr Sugimoto's inclusion in a 1985 listing of the Ten Greatest Inventors in Japan, which accolade may seem a trifle extravagant despite the obvious difficulties of designing a practical typewriter for that complex language.

Sugimoto graduated from the Osaka Telecommunications Technology Institute in 1900 at the age of eighteen, and after

259. Travis (Courtesy of Tom Fitzgerald)

working in the field of typography, began developing a machine capable of handling as many as possible of the thousands of characters in the Japanese language. He achieved this by designing a device which selected the desired characters from trays and transferred them to paper held in place around a cylindrical platen. It was of necessity a slow process since each individual character had to be replaced in its slot before another was selected by the same means, but this lack of speed was relatively unimportant at the time compared to the slow and laborious process of writing the language by hand. There are approximately 50,000 characters in the largest dictionary of the Japanese language; of these, between 1,500 and 3,500 are required for ordinary documents. This provides an insight into the task facing early inventors of Japanese typewriters and may help to explain why Sugimoto's apparently primitive invention—consisting essentially of a character storage section moving right and left, a printing section moving back and forth, and a cylindrical platen—should have become the prototype of the modern machine for that language, earning quite intemperate accolades for its inventor including a Blue Ribbon Medal in 1953 and a Fourth Class Order of the Rising Sun, as late as 1965.

Japanese word processors, of course, have now made all this obsolete. Illustration no. **(251)** is the original prototype, circa

260. Typatune (Courtesy of This Olde Office, Cathedral City, California)

1915; no. **(252)** from a 1931 newspaper shows that original instrument in use.

SUN

Lee S. Burridge and Newman Marshman of New York designed the first of several linear index machines of virtually identical specifications which were patented and manufactured at around the same time. Their patent, granted 1885, called for a sliding index at right angles to the platen. A small grip, attached to the index, served to select the desired letter on a scale, with slots for alignment, and pressing down on the grip brought the index in contact with the transverse platen below. Roller inking. The manufacturer was the Sun Typewriter Co. of New York. Other machines which were mechanically identical if aesthetically different included International, New American, and Odell.

Early 'dog-bone' model (so called because of the shape of the base) fetches mid-range four figure prices, later models tend to be in the high three to low four figures. **(256)**

SUN STANDARD

Another of Lee S. Burridge's patents, dating from 1899, protected a small front stroke machine with three row keyboard and double shift which featured a novel inking device. Manufactured by the Sun Typewriter Company of New York from 1901 onwards, the type bars were inked by a roller on their way to the platen, and the roller itself made contact with a larger roller which acted as a booster, keeping the small roller constantly inked. The system was understandably suspect even if it might be considered better than a single roller on its own,

and later models accepted the inevitable and switched to ribbons. Nine models were manufactured over the years, including one with four row keyboard in 1907. It was marketed in different countries under a variety of names which included Carlem, Leframa, New Sun, Nova, and Star.

Model 2 was the first of this machine to appear—there was no No. 1 and it must be assumed that Burridge considered the linear index Sun to be his first model, despite the fact that it was a totally different design.

Good value at up to $500-600 or so. **(8)**

TACHEOGRAFO

Manlio Marzetti of Milan designed this compact and elegant shorthand machine in 1904. It sported eleven keys to be used individually or in combinations.

TACHIGRAFO MUSICALE

An Italian by the name of Angelo Tessaro was responsible for the first musical typewriter to be manufactured; the year was 1887.

TACHOTYPE

Shorthand machine with twenty-seven keys printing on paper tape, manufactured in Holland and Germany shortly before World War II. Speeds of 250 syllables per minute were claimed for the instrument.

No more than three figures.

TAURUS

A quite unique Italian machine, manufactured by Torrani & Co. of Milano in 1908, the Taurus represents perhaps the worst misnomer in typewriter history, for it is nothing more than a miniature type wheel with a circular index in a case made to resemble a large pocket watch both in shape and size. Pressing the pendant button brings the narrow paper tape into contact with the type wheel. Inking by roller. Diameter: 7 cm.

Exceedingly rare, not surprisingly, so the value is likely to be in the mid-four figure range. **(56, 257)**

TIP-TIP

The Mignon was resurrected in this rather cheap and nasty format, with the type index moved from the left to the right-hand side of the machine. Ing Franz Hubl of Varnsdorf in Czechoslovakia manufactured it in 1936 and a total of 5,000 were said to have been built, although an example previously in the author's collection had a higher serial number.

Flimsier than the Mignon, but prices tend to reflect relative rarity and are mostly in the $200-400 bracket, although a few maverick exceptions have been recorded. **(258)**

TITANIA

Conventional German upright introduced in 1910 by A. G. Mix & Genest of Berlin, later Titania Mix & Genest Schreibmaschinen GmbH, then in 1913 by Titania Schreibmaschinen GmbH and finally in 1918 by Titania-Werke GmbH of Berlin. Several similar models. The design featured type bars individually mounted in tiny ball bearings.

A hundred or two ought to be more than enough.

TORPEDO

German upright of conventional design developed from the Hassia and manufactured from 1907 first by Weilwerke A. G. and then by Torpedo-Werke A. G., both of Frankfurt. Numerous models over the succeeding decades, including portables. Manufacturer was controlled by Remington from 1932.

A hundred or so is ample.

TRAVIS

William H. Travis patented this type wheel machine with four row keyboard from 1896 on, and the Philadelphia Typewriter Co. of Philadelphia, Pennsylvania, manufactured it briefly in 1905.

A very rare machine, this one, so somewhere in the $5,000-6,000 bracket for sure, and possibly considerably more. **(259)**

TRIPLE TYPEWRITER

This remarkable machine was built by a company in Cincinnati out of three Smith Model 2 double keyboard typewriters using only the type baskets and carriages of the rear two, attached to and operated from the single keyboard of the complete front machine. The idea was to construct a typewriter which would type three originals in one operation, the machine being designed to type on cards through which carbon paper would not otherwise have produced satisfactory copies. The special cards designed for the machine were fed from

261. Universal Crandall (Courtesy of DeWitt Historical Society, Ithaca, New York)

the first through the second onto the third carriage so that each was correctly aligned. It may justifiably be assumed that splints and plaster casts for the typists' fingers were likely to be required as optional extras…

If the vendor delivers it to your house, offer him a low three figure sum, despite the fact that the machine is rare. If you have to deliver it yourself, ask *him* to pay *you* the same amount for clearing it off his site for him.

TRIUMPH

Triumphwerke Nürnberg A. G. were responsible for this machine after their acquisition of the Norica works in 1909 and their remodelling of that unsuccessful design into a conventional upright. The Triumph was thus very much a German machine although the manufacturing company was initially a subsidiary of the British bicycle and motorcycle maker of the same name.

Cross-pollination with Adler typewriter interests dates back to the 1920s although the complete amalgamation of the two companies is a relatively recent development.

$50? $100? Somewhere in between, perhaps.

TRIUMPH VISIBLE

The Visible Typewriter Manufacturing Company of Kenosha, Wisconsin, briefly produced this oblique front stroke four row keyboard machine in 1907. Also sold under the names Imperial, Imperial Visible, and Triumph Perfect Visible. In 1905 it appeared in slightly modified appearance as the Burnett.

Rare…and when it comes to value, we all know what that means…**(94)**

TYPATUNE

1930s (or so) toy which types tunes as the keys are struck. Comes complete with a booklet of tunes, too. Good examples are already up in the three figures, these days. **(260)**

TYPEMUSIC

A Royal portable converted to print musical scores, with a lever on the right to raise the carriage to the required space or line.

A nice addition in the three figure bracket.

TYPE WRITER

Sometimes hyphenated, sometimes not, Type Writer was the name by which the Sholes and Glidden typewriter was known from its first appearance early in 1874 until it graduated into the Remington Model 2 some years later. The history and development of this machine is fully described and explained under the entries for Remington and Sholes and Glidden.

262. Underwood Standard portable (Courtesy of Tom Fitzgerald)

263. Velograph (Courtesy of Bernard Williams)

Initially mounted on a modified Remington sewing machine table and fitted with treadle carriage return, then removed from the table and fitted with side handle carriage return, finally with return lever fitted to carriage itself, the Type Writer in its final form, stripped of its external carriage return hardware, stripped of its front, rear and top covers, stripped of its copy holder, designated successively 'Improved' and finally 'Perfected,' was being offered in 1877 by W. O. Wyckoff of Ithaca, New York, the retail agent, in five versions: No. 1: Upper case only; No. 2: upper and lower case; No. 3: wide carriage model; No. 4: fitted in a 'noiseless' case (yes, that is what they said!); and No. 5: fitted in a 'portable' case (idem!) weighing a mere 20 lbs. This last model was of considerable interest to the growing army of young lady 'typewriters' who at the time were expected to turn up for work with their own Type Writers.

See corresponding entries under Sholes and Glidden, Improved Type Writer, and Perfected Type Writer.

TYPOGRAPH

A beautiful little instrument and only the second typewriter ever to be manufactured was this 1850 radial type plunger machine for the blind, designed by G. A. Hughes. Just how many were made is not known; the one illustrated bears the serial number 52. Circular index, with the letters in relief to permit them to be selected by feel; a corresponding circle of plungers, the desired one of which was pressed down onto the flat paper table beneath by means of a lever. The index and plunger assembly were fitted to a ratchet wheel which travelled along a spiral groove cut into a transverse rod so that each depression of the printing lever moved the assembly to the next space. Line spacing was performed by turning a thumb screw at the front of the machine through one complete revolution. Initially designed to print embossed letters which could be made visible by means of carbon paper.

Price? Who knows? If you see one for sale for less than a five figure sum, pay the vendor his money, nail his foot to the floor so that he cannot change his mind and chase after you, and run for your life! **(2, 44, 176)**

TYPOGRAPHER

John Jones of New York was responsible for this important machine—important not so much for the design itself (although it was good) as for the fact that it was only the third one ever to actually reach manufacture. And 'manufactured' it certainly was, albeit on a limited scale—no one-off job, this, for 130 machines were either completely or partially assembled when the factory with all its contents was destroyed by fire.

264. Velograph (Courtesy of Sotheby's, London)

Sometimes called more fully 'Mechanical Typographer,' the design was no worse than most and better than many similar machines which were to follow in the course of the next thirty-odd years after its introduction in 1852, the same year in which it was patented. It consists of a horizontal type wheel beneath a circular index, inking by roller, and typing on paper around a short platen of large diameter, complete with escapement and tension supplied by a spiral spring, the paper advance and platen return performed automatically at the end of each line. And as if all this were not enough, the machine offered full differential spacing! Quite a brilliant combination, considering the early date.

The fire wiped out Jones' first effort, but he tried again in 1856 when he was granted another patent for an improved Domestic Writing Machine of similar design to the first except that a single handle was now used to turn the type wheel and bring it down onto the paper, not separate levers as on the earlier machine. The cylindrical platen was replaced by an endless belt; upper as well as lower case was now offered and the differential spacing feature was retained. This time, however, the instrument was not manufactured.

Bank managers or possibly crime might be called for if a Typographer were to appear on the market, so brace yourself for a five figure hit.

UNDA

Austrian front stroke upright machine with three row keyboard developed but not manufactured prior to World War I by Ing. Endemann who had earlier been involved in the production of the Albus. F. A. Bechmann of Vienna finally succeeded in placing the machine on the market in 1921.

Zzzzzzzz…Austrians may be induced into paying funny money for one, although the rest of us would be unlikely to part with more than a high two or low three figure mark.

UNDERWOOD

Franz Xaver Wagner of New York was responsible for arguably the finest typewriter ever created, and that lavish and extravagant claim is advisedly made. Born in Germany in 1837, he emigrated to the United States in 1864 and was soon applying his mechanical skill to the development of the Sholes and Glidden Type Writer in the Remington factory. He helped Yost with the Caligraph and the Densmore brothers with the machine to which they gave their name. He held patents for machines of all kinds, employing just about every imaginable system, from segmental combs to linear index.

265. Victor

It was all experience which stood him in good stead, for when he ultimately filed a patent in 1893 for a visible front stroke upright machine with four row keyboard, the design embodied everything he had ever learnt. The Wagner Typewriter Company was formed to develop and manufacture the design but was soon bought up by John T. Underwood, who, along with his father, owned a successful typewriter ribbon and carbon supply company. The Underwood Typewriter Company was incorporated in 1895 with premises in New York, and the first prototypes were manufactured. In 1896, some fifty machines were produced; 286 more in 1897; 2,167 in 1898, and from then on, increases in production figures were exponential. The design requires little description for it was destined to become the prototype for future generations of typewriters the world over. Remington and others who were successfully manufacturing non-visible machines were soon fighting losing battles and, one by one, redesigning their products along Underwood lines, a fact lamented principally by all succeeding generations of typewriter collectors. The age of standardization had indeed dawned.

The Model 5 which was manufactured literally in its millions was introduced in 1900, previous models which are rare being distinguishable mainly by minor features such as length of platen. Production of the Model 5 continued for some thirty years. It was electrified, it was sound-proofed, it was enhanced with tabulators and it was partially converted for accounting and other special purposes. But throughout this period the essential design remained unchanged.

A line of portables was introduced in 1919 first as three row, then in 1926 as four row keyboard models. Early models in pristine condition might fetch a decent price these days, despite the ubiquity of the machine, but generally a hundred or two ought to be more than enough. **(75, 116 [r], 262)**

UNION

A conventional upright by this name was briefly manufactured in Germany in 1908 before being renamed Dea, since the name 'Union' was already protected by the Union Typewriter Company, the trust formed in 1893 by Remington, Yost, Smith Premier, Caligraph, and Densmore (and later Monarch) to control the typewriter market by fixing prices and stifling competition.

Not a commonly found name, but the machine is a snore, so if the price is high, so is the buyer!

URANIA

Conventional upright manufactured in numerous similar models from 1909 by Clemens Müller A. G. of Dresden. The same manufacturer was also responsible for the Perkeo folding portable which later became a conventional portable marketed as the Urania Piccola from 1925 and Klein-Urania in 1935.

The Urania was sold in various parts of the world under the names Concordia Gallia, Haddad, Merkur and Ujlaki; the Klein-

266. Victor 2 (Courtesy of This Olde Office, Cathedral City, California)

267. Victor 10 (Courtesy of This Olde Office, Cathedral City, California)

Urania as Aristokrat, Concordia, Merkur, and Regent. A model incorporating an adding and subtracting machine was introduced in 1920 under the name Urania-Vega.

See above entry, except that in this case the name is also common, so if you find yourself lugging it home, blame the hole in the ozone layer.

VARI-TYPER

The final incarnation of the Hammond typewriter, successfully adapted to the printing trade by the manufacturer when production of all other models had ceased. Introduced in 1927, it offered type shuttles for fifty-five languages and 300 type styles. Originally manufactured by the Hammond Typewriter Co. itself, from 1931 by Vari-Typer Inc. of New York, and finally by Ralph C. Coxhead Corp. of the same city.

Prices in the mid hundreds are not uncommon. **(171)**

VELOGRAPH

Swiss machine patented in 1886 by Adolphe Prosper Eggis. The design featured a circular index beneath which was rubber type inking by pad. Two indicator knobs were used for letter selection for upper and lower case respectively. Rymtowtt-Prince & Cie, Geneva manufactured it. In 1887 a modified design was introduced which featured a single central knob for letter selection and typing. A version of the machine designed as a cipher device, based on monalphabetic substitution, was called Eggis. A linear index model not dissimilar to the Odell printing upper and lower case was marketed in 1889 and also named Commerciale.

This beautiful and desirable instrument is likely to deplete your finances to the tune of $6,000-8,000. **(263, 264)**

VICTOR (1)

F. D. Taylor and J. A. White of Hartford, Connecticut, were responsible for this hybrid type wheel machine using a semi-circular index, produced by Tilton Manufacturing Co. of Boston, Massachusetts. Rubber type, pressed against the paper not by the indicator arm but by a separate lever. Pad inking. The design was protected by several patents including a British one in 1889 and manufactured some years later, the precise date uncertain. Few were made.

Don't even blink a $3,000-4,000 price tag. **(265)**

VICTOR (2)

Front stroke upright of conventional format in several similar models manufactured from 1907 by the Victor Typewriter Co. of New York in the former Franklin factory. The machine enjoyed considerable success and was widely exported to Europe. Sold in Germany under the name Diktator.

Expect to pay anything up to high two or even three figures (just), if the Franklin factor induces you to buy at all. **(266, 267)**

VICTOR (3)

Victor Adding Machine Corp. of Chicago briefly marketed this conventional four row keyboard machine from 1927.

Zzzzzzzz…Cross polination with calculator enthusiasts is likely to result in a three figure price tag.

268. Victoria **(l)**, Practical **(c)**, Junior **(r)** (Courtesy of Bernard Williams)

VIROCYL

Small but pleasant and well-made toy, employing ribbon, carriage, and type wheel.

Likely to set you back $500 or so. **(269)**

VIROTYP

A somewhat improbable design which nevertheless enjoyed considerable popularity during its relatively short lifetime and which appeared in a number of different but related models. A Frenchman by the name of H. Viry designed it in 1914, initially for use during the First World War in circumstances, such as in the trenches, when no surface was available on which to rest a conventional machine. Operation entailed hooking the little finger of the left hand through a strap on the left of the machine and holding the right side by sliding two curved hooks between the index and middle fingers of the right hand. That left the thumb and index finger of the left hand free to turn the indicator knob and the thumb of the right hand free to press the printing key. Type was on individual plungers be-

neath the circular index. Two rollers beneath the type held paper in place.

Various models were produced. Initially fitted with two straps for attaching to a forearm for mobile operation such as on horse-back, it was later fitted to a base for table-top use, this model designated 'G,' and an even larger table version was also manufactured. Machines previously in the author's collection alone have made it possible to track serial numbers up to almost 10,000, which, given the late date and the strangeness of the design, is quite a tribute to M. Viry.

This is a popular and spectacular little machine, and excellent value at prices up to $1,000. **(270, 271)**

VISIGRAPH

Conventional front stroke with four row keyboard designed by typewriter pioneer Charles Spiro in 1910. Spiro's inventive efforts spanned four decades and dozens of patents, including such great machines as the Columbia type wheel and the Bar-Lock. The Visigraph was manufactured first by the Visigraph

269. Virocyl (Courtesy of Bernard Williams)

270. Virotyp, portable model **(c)**, Merritt **(l & r)** (Courtesy of Sotheby's London)

271. Virotyp, table model (Courtesy of Sotheby's London)

272. Volksschreibmaschine

Typewriter Co. of New York and later by the C. Spiro Manufacturing Co. of the same city. The business was bought up by the Federal Typewriter Co., also of New York, in 1919 and briefly marketed under the name Federal.

$100-150 ought to do it.

VITTORIA

Italian conventional upright machine introduced in 1923 by Fratelli Bertarelli of Milano.

Probably of interest to few people outside Bertarelli family circles; however, you might even find yourself paying fifty bucks or so, if mamma has softened you up with a good pasta dinner.

VOLKSSCHREIBMASCHINE

A vertical type wheel geared to an indicator arm which selected letters from an oblique circular index, manufactured in Germany by Friedrich Rehmann of Karlsruhe in 1898. Paper was held on the flat base and printing was performed by pressing a lever on the left which brought the type wheel down onto the paper. Inking by roller. A somewhat improved model called Diskret, introduced in 1899, was offered as a cipher machine by the simple expedient of changing the position of the type wheel relative to the index, thereby printing a simple substitution code which was quite worthless and unlikely to have proved challenging to anyone above primary school level. As collector's items, however, the Volksschreibmaschine (also called Volks) and the Diskret (also called Discreet) are rare and highly prized. Few were made and, of course, fewer have survived.

This is one of those machines which tend to give collectors (particularly German collectors) a glazed look in their eyes, and when that happens, it can only mean one thing: a price tag in the four to five figure bracket. **(272)**

WAGNER & SCHNEIDER

Horizontal circular index machines for the blind, offered in several models by this Swiss manufacturer in 1888. Permutations of printing and embossing were offered to permit the sighted to correspond with the blind, the blind with the sighted, and the blind with the blind.

Prepare yourself for three to four figure prices.

WAVERLEY

If you can accept the concept that a patient can die during a successful operation, then you will have no trouble understand-

273. Williams 3 (Courtesy of Christie's South Kensington, London)

ing that the Waverley 'was a magnificent piece of work but the Company did not last long.'[62] But 'magnificent piece of work' it truly was, this British down stroke machine patented by Edward Smith Higgins and Henry Charles Jenkins in 1889 and manufactured briefly by the Waverley Typewriter Company in London before the company packed up in 1897.

Four row keyboard with the vertical type bars in an arc behind the platen. Initially inking by roller but modified to ribbon almost immediately after its introduction. It embodied many interesting and progressive features including terminal and differential spacing as well as a novel mechanism which prevented accidental partial shift.

Start the bidding at maybe $8,000 or so, and keep going…keep going…keep going… **(80)**

WESTPHALIA

E. W. Brackelsberg was granted a German patent in 1884 for an interesting linear index device with differential spacing of which a small number is said to have been manufactured. Also sometimes referred to under the name of the inventor.

Too rare to call, which can only mean serious money.

WILLIAMS

John Newton Williams of New York spent fifteen years from his first 1875 patent developing the remarkable 'grasshopper' machine to which he gave his name. The Domestic Sewing Machine Co. first marketed the product in 1891 followed shortly afterwards by the Williams Typewriter Co. of Derby, Connecticut. Williams is reported to have worked on his design at the same factory where the ill-fated Fitch was being made and some features of his machine are said to have been copied from the Fitch. The reference may well be related to the basket in which the paper was curled.

The significant feature of the Williams design is the centrally-located platen with horizontal type bars arranged in sectors front and rear. Three row keyboard. Type bars rest face downwards on inking pads; on pressing a key, the corresponding type bar lifts off its pad and performs a neat hop onto the platen. The first examples had curved three row keyboards but a straight keyboard was introduced almost immediately and from serial numbers on surviving machines there is con-

siderable evidence that for some time curved and straight keyboard models were manufactured concurrently before the curved version was dropped. The three row Model 3 was slightly modified to accommodate a wide-carriage. The four row keyboard Model Four was introduced in 1900. Similar Models 5 and 6 followed.

One of the characteristics of the Williams which may well have been Fitch-inspired was the open framework into which the blank page was curled and inserted into the machine in front of the platen, and the one behind the platen into which the typewritten page was progressively fed. It was a clumsy arrangement but with type bars front and rear hopping onto the platen, there was nowhere else for the paper to go. Ultimately, faced with growing competition from conventional machines, the design was doomed and the company went into receivership in 1909 when it was taken over by the Secor Typewriter Co. which produced its own front stroke upright.

Not a rare machine but a particularly desirable one. Curved keyboard examples are relatively rare and have been known to fetch $4,000-5,000 but later models are very much cheaper and sometimes struggle to reach $1,000 unless they are particularly good examples. **(20, 54, 76, 273, 274)**

WOODSTOCK

Conventional upright manufactured from 1914 by the Woodstock Typewriter Co. of Woodstock, Illinois. The company was previously known as the Emerson Typewriter Co. before being bought up by Roebuck of Sears Roebuck who changed its name first to Roebuck Typewriter Co. and eventually to Woodstock. The first Woodstock was a cheaply made Oliver offered for sale by mail order in 1898-9 but a mere nineteen examples were apparently all that were ever manufactured. The subsequent front stroke Woodstock was designated model no. 3, followed by a model 4. A popular machine, produced for many decades in numerous successive models, including electrics and portables.

Generally very cheap, but selected ones have been known to make even up to three figures. **(98)**

WORLD

John Becker of Boston was responsible for this simple swinging sector machine using rubber type which he patented in 1886. The design featured a semi-circular carrier to which was attached an indicator arm for selecting letters from a curved index. Printing was performed by lowering a bar which ran the length of the machine, with holes in the type sector providing positive alignment. Pad inking. Manufacturer was first World Type Writer Co. of Maine to whom patent was assigned, then Pope Manufacturing Co. of Boston and finally Typewriter Improvement Co. of the same city. Model One printed upper case only; Model Two of 1890 was the upper and lower case version. Also marketed as Boston and, later, as New World.

Despite the primitive design of this and other comparable indicator machines, they were nevertheless quite popular in their day. Serial numbers of surviving Worlds have been traced into five figures.

Not particularly rare, but prices for this intriguing instrument range from somewhere around $600 all the way to $1,000. **(275, 276)**

WRIGHT & NIGRON

Early printing telegraph machine using a type wheel, manufactured in small numbers around 1907. An example previously in the author's collection bore serial number 221. John Edward Wright held patents dating back to 1889 but the one protecting the printing telegraph was dated in 1892.

Rare, but printing telegraphs tend not to be particularly sought after. Price in the mid-three figures.

274. Williams 4 (Courtesy of Bernard Williams)

Scientific American

275. World[83]

XCEL

Syllable typewriter which used a conventional four row keyboard plus a further row of keys for syllables and short words, this machine represents the second attempt by Wesley Henry Bennington to market a syllable typewriter, the first one being the machine to which he gave his name. The Xcel Typewriter Corp. of New York introduced the machine in 1922 but with limited success.

Rare, so the price will reflect this.

YETMAN

The Yetman Transmitting Typewriter manufactured by the Yetman Typewriter Co. of New York was basically a front stroke upright with a four row keyboard adapted for telegraphy so that messages could be simultaneously typed and transmitted. A Monarch Visible provided the basis on which the machine was built in 1903 by Charles E. Yetman of Washington, D. C. The company went into liquidation in 1909 and its assets sold to J. L. Smith of Philadelphia, Pennsylvania, who briefly marketed the machine as the Smith Visible.

Rare but not necessarily priceless. $2,000-3,000 ought to be about right. (**278**)

YOST

George Washington Newton Yost first appeared in typewriter history as the man principally responsible for selling Remingtons on the idea of manufacturing the Sholes and Glidden back in 1873. For his efforts he was cut in on the

276. World (Courtesy of Sandy Sellers)

277. Write Easy

action, and from then on he was never again able to distance himself from some aspect or other of the typewriter. His involvement as a partner in the ownership of the machine proved of little financial value, and the retail partnerships first of Densmore, Yost and Co. and then Locke, Yost and Bates were equally unprofitable.

Left to his own lavish devices he tended to be more successful, but by then the groundwork had already been done and the typewriter was a reality. His first independent venture was the Caligraph which enjoyed considerable success as the principal rival to Remington but Yost's involvement came to an end when he fell out with his associates and embarked on the new venture to which he gave his name.

The Yost was a double keyboard machine of the 'grasshopper' group, the type bars arranged in a circle beneath the printing point with the type face resting against an inking pad. By an action peculiar to this device, pressing a key caused the type bars to lift off the pad and perform a delicate turn before striking up through the central guide at the printing point. The design was the responsibility of Alex Davidson, Andrew Steiger, and Jacob Felbel whose patent, filed in 1887, was granted two years later. First manufactured in the Merritt factory in Springfield, Massachusetts, and then by the Yost Typewriter Co. of Bridgeport, Connecticut.

Model Four was the first appreciable departure from the original machine and was introduced in 1895 but all basic features were retained, as they were on the next model to appear, the Model Ten, which was of more substantial construction and which was introduced in 1902. Model Fifteen, 1908, finally succumbed to competition and appeared as a four row front stroke visible and the 1912 Model Twenty was virtually identical.

Some examples of the first model were also marketed as New Yost, and the Model 15 as the Model A.

Many examples of this intriguing design have survived. The earliest model is the most expensive at prices ranging up to

$1,000 or so; later grasshopper models drop vertiginously (in price, that is) and represent very good value for money. **(53, 60, 279)**

ZALSHO

Four row front stroke upright briefly manufactured in London by Zalmon, son of Christopher Latham Sholes. Better remembered for the Rem-Sho, which folded in 1909, Sholes then became involved in the production of a machine called Acme after which he transferred operations to London where he succeeded in interesting a group called Lawrence Manufacturing Co. in producing the machine in 1913, but the outbreak of war doomed the project to a premature demise two years later. He then re-crossed the Atlantic with the scheme, and the machine briefly re-appeared as the Z. G. Sholes, produced by the Standard Arms Manufacturing Co. of Wilmington, Delaware. The Sholes Standard Typewriter Co. was formed to handle marketing of the machine but Sholes' death in 1917 brought the whole project to an end.

Interesting only by association with the inventor, so the value is likely to be modest but speculative.

278. Yetman (Courtesy of Bernard Williams)

279. Yost (Courtesy of Bernard Williams)

BIBLIOGRAPHY

1. ADLER, M. *The Writing Machine*. London, 1973.
2. ALIPRANDI, G. *Giuseppe Ravizza, Inventore della Macchina da Scrivere. Novara*, 1931.
3. ALIPRANDI, G. *Cenni Storici della Macchina da Scrivere. Padova*, 1938.
4. ALIPRANDI, G. *Giuseppe Ravizza Attraverso le Pagine del Suo Diario*. Novara, 1942.
5. BACULO, L. *Manuale di Stenographie*. Napoli, 1849.
6. BAGGENSTOS, A. *Von der Bilderschrift zur Schreibmaschine*. Zürich, 1977.
7. BEECHING, W.A. *Century of the Typewriter*. London, 1974.
8. *Berlingske Tidende*, 1878.
9. BLIVEN, B. *The Wonderful Writing Machine*. New York, 1954.
10. *Bollettino della Accademia Italiana di Stenografia*, 1926, 1930, 1931, 1936.
11. British Association for the Advancement of Science, York Meeting, 1844.
12. BUDAN, E. *Le Macchine da Scrivere dal 1714 al 1900*. Milano, 1902.
13. *Bulletin de la Société d'Encouragement pour l'Industrie etc.* 1831, 1843, 1850, 1881.
14. BURGHAGENS BÜROFACH-BÜCHEREI. *Liste der Herstellungsdaten etc.* Hans Burghagen Verlag, Hamburg.
15. CAIZZI, B. *Gli Olivetti*. Torino, 1962.
16. CHAPUIS, A. *Histoire de la Boîte à Musique*. Lausanne, 1955.
17. CHAPUIS, A. & DROZ, E. *Automata*. Neuchâtel, 1958.
18. CHAPUIS, A. & DROZ, E. *Les Automates*. Neuchâtel, 1949.
19. CHAPUIS, A. & DROZ, E. *Les Automates des Jaquet-Droz*. Neuchâtel, 1951.
20. CHAPUIS, A. & GÉLIS, E. *Le Monde des Automates*. Neuchâtel, 1928.
21. COOKE, T. F. *Authorship of the Practical Electric Telegraph etc*. London, 1868.
22. CURRENT, R. *The Typewriter and the Men Who Made It*. Illinois, 1954.
23. DENMAN, R. P. Q. *Electrical Communication*. London, 1926.
24. *Design and Work*, 1877.
25. *Dinglers Politechnisches Journal*. 1822, 1872, 1881, 1883, 1887, 1888.
26. DUPONT, H. & CANET, L. F. *Les Machines à Écrire*. Paris, 1901.
27. DUPONT, H. & SÉNÉCHAL, G. *Les Machines à Écrire...Leur Évolution*. Limoges-Paris, 1906.
28. DUPONT, H. & SÉNÉCHAL, G. *Les Machines à Sténographier*. Limoges-Paris, 1907.
29. DYER, F. L. & MARTIN, T. C. *Edison, His Life and Inventions*. New York, 1910.
30. *English Mechanic and Mirror of Science*, 1866.
31. Exhibition of the Works of Industry of All Nations. *Official Catalogue...*London, 1851.
32. FAESSLER, F., GUYE, S. & DROZ, E. *Pierre Jaquet-Droz et son temps*. La Chaux-de-Fonds, 1971.
33. FINLAISON, J. *Some Remarkable Applications of the Electric Fluid...*London, 1843.
34. *Frankfurter Ober-Postamts-Zeitung*, 1831.
35. FREEBODY, J. W. *Telegraphy*. London, 1959.
36. GRANICHSTAEDTEN-CZERVA, R. *Peter Mitterhofer, Erfinder der Schreibmaschine*. Vienna, 1923.
37. GRETCHEN SACHSE et al. *The Spirit of Enterprise...*Ithaca, New York, 1977.
38. *'GRIP'S' Historical Souvenir of Groton*, 1899.
39. *Groton and Lansing Journal*, 1909.
40. HARCOURT-SMITH, S. *A Catalogue of the Various Clocks, etc.* Peiping, 1933.
41. HERKIMER COUNTY HISTORICAL SOCIETY. *The Story of the Typewriter, 1873-1923*. New York, 1923.
42. HERRL, G. *The Carl P. Dietz Collection of Typewriters*. Milwaukee, 1965.
43. *Historische Bürowelt & HBw Aktuell*. Köln, Germany.
44. HUBERT, P. G. *The Typewriter, its Growth and Uses*. 1888.
45. *Illustrierte Zeitung*, 1872.
46. *Inland Printer*, 1886.
47. *International Congress on Technology and Blindness, proceedings of...*New York, 1963.
48. International Exhibition 1862, London. Reports by the Juries.
49. *Ithaca Journal*, 1889, 1890.
50. JENKINS, H. C. *Cantor Lectures on Typewriting Machines*. Journal of the Society of Arts, 1894.
51. JONES, C. LEROY. *Typewriters Unlimited. History of the Typewriter*. Missouri, 1956.
52. KAHN, D. *The Codebreakers*. New York, 1967.
53. KARRASS, TH. *Telegraphen-und Fernsprech-Technik*. Berlin, 1909.
54. KRCAL, R. *1864-1964 Peter Mitterhofer und Seine Schreibmaschine*. Aachen, 1964.
55. *Kwartaalblad voor de Schrijfmachineverzamelaar*. Tilburg, Holland.
56. *La Lettura*, 1908; *Antiche Lettere Scritte a Macchina*. Umberto Dalari.
57. *La Nature*, France, 1881.
58. LAUFER, R. *Textes Presentés par... La Machine à Écrire Hier et Demain*. Paris, 1982.
59. *Leertaste*, Germany.
60. LIPPMAN, P. *American Typewriters*. New York, 1992.
61. *Magazin Aller Neuen Erfindungen*, 1808.
62. MARES, G. *The History of the Typewriter*. London, 1909.

63. MARTIN, E. (pseudonym of JOHANNES MEYER) *Die Schreibmaschine und ihre Entwicklungsgeschichte.* Pappenheim, 1949.
64. MASI, F. T., ed. by. *The Typewriter Legend.* New Jersey, 1985.
65. MATHEWS, W.S.B. *The Writing Machine.* Northwestern Christian Advocate, 1891.
66. *Mechanics Magazine,* 1832.
67. Milwaukee County Historical Society. *Dedication Pamphlet.*
68. MORELLI, D. *La Storia della Macchina per Scrivere.* Brescia, 1956.
69. MÜLLER, F. *Schreibmaschinen.* Berlin, 1900.
70. Museo Nazionale della Scienza e della Tecnica, Milano. *Commemorative Exhibition Publication,* 1955.
71. NOGUEIRA, J. C. de A. *Um Inventor Brasileiro.* São Paolo, 1934.
72. ODEN, C. V. *Evolution of the Typewriter.* 1917.
73. PERREGAUX, C. & PERROT, F-L. *Les Jaquet-Droz et Leschot.* Neuchâtel, 1916.
74. PROUDFOOT, W. B. *The Origin of Stencil Duplicating.* London, 1972.
75. QUAIFE, M. *Henry Roby's Story of the Invention of the Typewriter.* Wisconsin, 1925.
76. REES, A. *Cyclopædia.* London, 1819-1820.
77. *Reports of the Paris Exhibition,* 1867.
78. RIDDELL, A., ed.by. *Typewriter Art.* London Magazine Editions, 1975.
79. ROBLIN, J. *Louis Braille.* London.
80. ROCHEFORT-LUÇAY, M. D. *Les Machines à Écrire.* Paris, 1896.
81. ROUSSET, J. *Les Machines à Écrire.* Paris, 1911.
82. ROYAL SOCIETY, *London Philosophical Transactions,* No. 483, 1747.
83. *Scientific American* 1867, 1877, 1887, 1903.
84. SÉNÉCHAL, G. *Descriptions des Machines à Écrire Françaises.* Paris, 1903.
85. SHAFFNER, T. P. *The Telegraph Manual.* New York, 1859.
86. SPENCER, H. *Autobiography.* London, 1904.
87. *The Typewriter Exchange.* Arcadia, California.
88. TILGHMAN RICHARDS, G. *The History and Development of Typewriters.* London 1964.
89. TURNBULL, L. *The Electro-Magnetic Telegraph.* Philadelphia, 1853.
90. *Typewriter Times.* Luton, Bedforshire.
91. *Typewriter Topics Magazine.*
92. TYPEWRITER TOPICS. *History of the Typewriter.* Metropolitan Typewriter Co., Michigan, 1923.
93. WELLER, C. *The Early History of the Typewriter.* Indiana, 1921.
94. YAMADA, Hisao. 'Historical Study of Typewriters and Typing Methods.' *Journal of Information Processing,* Vol. 2 No. 4, Tokyo, 1980.
95. ZETZSCHE, K. E. *Geschichte der Elektrischen Telegraphie.* Berlin, 1877.

INDEX

Entries covering specific machines and inventions are listed below as fully as individual examples demand. However, general subjects common to all machines, such as inking methods, keyboards, carriages, type bars wheels and indices, and so on, are discussed in detail throughout the entire text and therefore individual generic entries for these features are listed below principally in order to establish definitions, systems, and categories. No attempt is made to list every single instance in which these features appear in the text.

Page numbers in **bold type** indicate illustrations.